SOUTH WALES COLLIERIE

VOLUME THREE

Valley, Vale, Coastal Collieries

From Ogwr to the Western Valleys of Afan, Neath, Dulais, Swansea, Aman, Loughor, Gwendraeth in Carmarthenshire and to Pembrokeshire in the far West of the South Wales Coalfield.

Pethau Disglaer/Shining Things

Words and music by Hawys Glyn James

1. Am bethau disglaer diolchwn ni,
   Adenydd cain yr aderyn du,
   Am ddisglaer rew ar ffenestri draw,
   Palmentydd gloyw a gwlyb gan law,
   Am ddisglaer sêr fel goleuadau'r ffair,
   Am fflamau tân a'u tafodau aur.

2. Am eira disglaer fel cwrlid gwyn
   A'r lloer fel arian uwchben y llyn,
   Mae lamp y glöwr a'i llachar bryd
   Yn goleuo'r ffordd fel llusern hud,
   Fel drych mae'r tô pan yn wlyb gan law,
   Ac adain gwylan yn fflachio draw.

3. Am donnau'r llyn sydd fel gemau claer
   Pan wena'r haul a'i belydrau aur,
   Am we pry copyn a'i berlau mân
   A dagrau'r gwlith ar y rhosyn glân,
   Am ymyl arian y cwmwl gwyn,
   A'r nant fel neidr yn gwau trwy'r glyn.

1. O Father, thank you for shining things,
   The sparkling sea and a blackbird's wings,
   For gleaming frost on the window pane
   And pavements shining and wet with rain,
   For glittering stars when nights are cold,
   For fire flames bold with their tongues of gold.

2. For glistening snow like a blanket white,
   The golden moon with its glowing light,
   And gleaming bright in the black coalmine
   Are miners' lamps as they boldly shine,
   For mirrored roofs on a rainy day,
   A seagull in sunlight, silver-grey.

3. For glinting rays of the sun that make
   A jewelled brooch of the shimmering lake,
   For morning dew on the shining leaves,
   A glistening web that the spider weaves,
   For silver-rimmed clouds with sun behind,
   And gleaming rivers that twist and wind.

South Wales Collieries

VOLUME THREE

Valley, Vale, Coastal Collieries

From Ogwr to the Western Valleys of Afan, Neath, Dulais, Swansea, Aman, Loughor,
Gwendraeth in Carmarthenshire and to Pembrokeshire in the far West of the South Wales
Coalfield.

David Owen

TEMPUS

Acknowledgements

Thank you all for the wonderful stories, songs, poems, drawings and photographs of the South Wales collieries that have been given to me by the people from the mining villages of South Wales.

These have come from the early days of the coal industry through to the new millennium. I dedicate my book to the people of South Wales, the land of song, and in memory of all the miners who worked at the collieries.

I sincerely thank everyone for their kindness and help.

David Owen
Author and Archivist

Cydnabyddiaeth

Diolch am yr holl storïau, caneuon, cerddi, darluniau a ffotograffau aruthrol o Faes Glo De Cymru, sydd wedi eu cynnig i mi gan bobl pentrefi glofaol De Cymru.

Mae'r cyfraniadau yma yn dod o ddyddiau cynnar y diwydiant glo trwyddo i'r milflwydd newydd. Rwy'n cyflwyno'r llyfr yma i bobl Gwlad y Gân De Cymru er cof am y glowyr a wethiodd yn y pyllau glo.

Rwy'n diolch yn ddidwyll i bawb am eu caredigrwydd a cymorth.

David Owen
Awdur ac Archifydd

First published 2002
Copyright © David Owen, 2002

Tempus Publishing Limited
The Mill, Brimscombe Port,
Stroud, Gloucestershire, GL5 2QG
www.tempus-publishing.com

ISBN 0 7524 2775 X

TYPESETTING AND ORIGINATION BY
Tempus Publishing Limited
PRINTED IN GREAT BRITAIN BY
Midway Colour Print, Wiltshire

Contents

Preface

Black Gold – Aur Du, The Story of Coal – Hanes Glo

The Promised Land: 200 million years ago the coalfield was compressed by an immense force from the north and south. These and other breaks in the strata raised or lowered the blocks of rocks between them. These violent movements caused other more complicated breaks in the strata; beds of rock and coal seams broke, one sliding on top of the other, with the result that double sections of coal were sometimes formed for short distances. Despite the complex mining problems presented by faults, contortions and thrusts, one hundred per cent of the output from the last remaining collieries that was cut and loaded by machines is a tribute to engineering and technological change.

The story of the South Wales collieries is a story of victory: victory of spirit over hardship, faith over misfortune. This special collection of photographs charts the great forces that have shaped the Valleys and their people: the force of toil, and sometimes greed, as men were pushed ever harder to hunt for the buried treasure that was Black Gold – Aur Du; the force of the faith as the chapels asserted their dominance in society; the force of progress as carts gave way to cars and trains and our streets grew old and were born again and most importantly the force of human dignity that has helped the Valleys triumph over adversity.

The photographs are real signposts in history. In this collection are a number of rare pictures depicting the great changes that occurred in the mining industry.

The Valleys exploded into life with the hewing of their vast resources of coal. What was the green and pleasant land was suddenly transformed into the most industrialised corner of Britain. The Valleys were brought to life by the thousands of people who came to live here. The coal rush attracted people from the rural West, the Border Counties, the North and nearer home, the ironworkers from South Wales. The dangers and hardship of mining, coupled with the lie of the land, ensured close-knit communities up and down the Valleys and a camaraderie as strong as blood bonds.

Coal was central to everybody's life: sixty per cent of the population were involved in the mining of it and many more involved in transport. The Valleys became legendary, as did the character of the people who lived there. For alongside the despair that must accompany an industry so vulnerable to tragedies, was the hope of the humour. The hope manifested itself in the vast number of chapels built and the fervour with which families took part in church life. There were the world-famous choirs, touching millions with their soul moving renditions.

The Price of Coal

It's a price that, for many Valley families, is counted in terms of human life and suffering. Pit disasters were all too familiar as part of everyday life. It was all too often a case of what miners' leader and former Lewis Merthyr miner A.J. (Arthur James) Cook referred to as an intensification of production at the expense of the safety of human life.

Today the monuments erected in the Valleys are a tribute to those who were killed in the pits as a result of accidents and explosions. The number of men who have lost their lives mining is comparable to the number of local people killed in the war.

One of the Valleys' greatest assets today is their splendid surroundings and the visitor has many fine walks to choose from, to explore and discover the secrets of this fascinating area. The bustling Valleys, gouged out by glaciers long ago, are set deep amongst miles of rolling mountains and forests. The area has many features of scenic, geological, historical and ecological interest and still contains many traces of the area's past.

Well over 3,000 million tons of coal have been extracted from the Valleys of South Wales since intensive mining began nearly two centuries ago.

David Owen
Author and Archivist

Foreword

The coal reserves formed millions of years ago have been extracted from deep below the surface of this land. Meanwhile, on the surface, now that the coalmining has come to the end of an era, the Valleys are being transformed, as work to remove the remaining scars left by the coal industry goes on through reclamation projects. New land formations, lakes, large tree-planting projects and the development of new areas for enjoyment of the countryside will ensure that the process of the changing of the shape and features of the Valleys is set to continue for the benefit of residents and visitors alike. This book will illustrate the importance of remembering how and why the Valleys developed as they did. Follow one of the leafy country lanes criss-crossing this lush green swathe of rolling hills and wooded valleys which link delightful villages waiting for you to discover them.

The past is rich with our inheritance and the photographs within this book by author David Owen are reminders of this proud heritage. This book presents a pictorial history of mining valleys and vales at Ogwr, Afan, Neath, Dulais, Swansea, Aman, Loughor, Gwendraeth in Carmarthenshire and Pembrokeshire in the far west of the South Wales Coalfield.

Alexandra Bowen
Director of Valley and Vale Community Arts, Betws

Rhagair

Mae'r cronfeydd o lo a grëwyd miliynau o flynyddoedd yn ôl wedi eu cloddio o ddyfnder mawr islaw wyneb y tir. Mae'r gweithfeydd glo wedi dod i ben erbyn hyn, ac mae'r cymoedd wedi eu trawsnewid wrth i waith fynd ymlaen i gael gwared o'r hen greithiau trwy brosiectau adennill tir. Mae'r tirwedd newydd, llynnoedd, prosiectau eang i blannu coed a datblygiad ardaloedd newydd i fwynhau cefn gwlad, yn golygu y bydd yr ardal er budd y brodorion a'r ymwelwyr yn yr un modd. Mae'r llyfr yma yn dangos y pwysigrwydd o gofio sut y datblygodd y cymoedd fel y maent. Dilynwch y llwybrau gwyrdd sy'n mynd yn ôl ac ymlaen ar draws y bryniau a dyffrynnoedd coediog, sy'n cysylltu y pentrefi hyfryd sy'n aros i'w darganfod gennych chi.

Mae'r awdur David Owen yn ein hatgoffa, gyda balchder, yn nhermau treftadaeth o'n gorffennol cyfoethog. Mae'r llyfr yn cynnig lluniau hanesyddol o'r cymoedd a chymoedd Ogwr, Afan, Nedd, Dulais, Tawe, Aman, Llwchwr, Gwendraeth yn Siroedd Caerfyrddin a Phenfro yn Ne Orllewin y Maes Glo.

Alexandra Bowen
Cyfarwyddwr Celfyddydau Cymunedol y Cwm a'r Cymoedd, Betws.

Introduction

David Owen's excellent volumes of photographs on the South Wales Coalfield have made him one of the great chroniclers of our rapidly changing valley landscapes. The photographs he has carefully collected and reproduced, the detailed historical commentary he has written and the masses of statistics he has carefully compiled all add up to a monumental achievement.

For me the great value of this present volume, as with all those which have preceded it, is the way in which the photographs are so revealing of the character of the people and the communities they created. At one and the same time, there is something unique and shared across all the Valleys' history: qualities of solidarity, fellowship, cooperation, sharing. But at the same time each valley, and certainly each community, has its own character. The Gwendraeth is not the Rhondda, Cwmgiedd is not Tredegar. These volumes help us to understand these differences.

In my life I have lived in three valleys: the Dulais Valley, where I was born and where I have spent most of my life; the Swansea Valley, where my grandparents' roots are and where I lived with my young family in the 1970s; and now more recently the Afan Valley, at Cwmafan, where I live in the Aberavon constituency.

In each of these valleys, there is a special enduring character, despite the vast economic changes which have come with the disappearance of coalmining. And I suppose that endurance is best displayed at Cwmafan where the memory of its two greatest sons, William Abraham (Mabon) and Richard Burton, is as strong today as it has ever been.

In their different but complementary ways they personify the old coalfield at its best: proud of their community roots, using their brilliant talents in the service of others and defining their valleys and their Wales as worldwide.

I hope there will be many more volumes of this kind produced by David Owen.

Dr Hywel Francis MP,
Cwmafan,
St David's Day 2002

One

Ogwr to Loughor Valley in the South Wales Coalfield

This great coalfield is assumed by various authorities to be approximately 1,000 square miles, which were distributed as follows: Glamorganshire, 518 square miles; Breconshire, 74 square miles; Carmarthenshire, 228 square miles; Pembrokeshire, 76 square miles; and Monmouthshire, 104 square miles.

Of the above, nearly 846 square miles are exposed, about 153 square miles lie beneath the sea and about 1 square mile is covered by newer formations.

We move on with our journey from the Glamorganshire central Valleys of Cynon, Rhondda and Ely (South Wales Collieries Volume One and Volume Two) and, with this Volume Three, from Ogwr to the western Valleys of Afan, Neath, Dulais, Tawe, Aman, Loughor, Gwendraeth in Carmarthenshire and to Pembrokeshire in the far west of the South Wales Coalfield.

The dramatic shapes and features of the Ogwr Valleys landscape that are before us today have been determined both by the impact of the forces of nature over millennia and the action of humans over centuries.

The formation of coal millions of years ago, its relatively recent discovery and development as a source of power and the carving of the valleys by water and ice have been the major factors in dictating how the Ogwr Valleys have been formed.

The story begins around 340 million years ago when the land we now call Wales was much nearer to the Earth's equator than it is now. Only an area of land in south and mid-Wales stretching to the Midlands was above water, and the rest of the country was a shallow warm sea.

In this area, known as St George's land, dense tropical forests grew and large river deltas flowed into the sea.

During this time, frequent changes in sea level drowned the huge forests, leaving them underwater on the seabed. As the water remained high, a layer of mud settled on the forest and over many years the trees were crushed by the tremendous weight of water and mud, transforming them first into a layer of peat and finally into a coal seam.

As sea levels fell back again, the uncovered layer of mud became rock and soil providing the environment for new forests to grow.

This process was repeated many times throughout the 65 million years of the carboniferous period, with each new forest drowned and buried by the sea to form yet more seams of coal.

As land masses around the world broke up to create the continents familiar to us now, Wales was no longer the tropical environment of dense forest and river deltas. It was now affected by a more northerly climate and the exposed land was free from the changes in sea level, which had played their part in forming the coal seams lying underground.

What we now call the Ogwr Valleys was originally part of the high plateau called the Blaenau which extended into Powys and covered the north of Glamorgan, with the low land or 'Bro' (now the Vale of Glamorgan) to the south.

The valley rivers were responsible for beginning the change from plateau to valley. Originating as springs on the plateau and flowing down to the sea, the water eroded and undercut the softer rocks on the sides and the beds over which it flowed, avoiding the harder rocks it was unable to wear down and formed channels down to the sea.

This process meant that over thousands of years the power of the rivers formed V-shaped valleys, as it cut into the underling rocks and wore down the valley sides.

The second major force in shaping the Valleys was that of slow moving ice in the form of a huge glacier. The Ice Ages, which gripped the earth many thousands of years ago, brought the tremendous power of a glacier to the Ogwr Valleys. As this massive ice pack moved slowly down the existing river valleys, pieces of bedrock became frozen into its base and were then plucked out of the ground as the glacier moved on. This rock debris caused further erosion by grinding and gouging the land beneath as it moved. The size and power of the ice pack wearing the land down over thousands of years meant that the Ogwr Valleys became the wider U-shape that can be seen today. Hollows where ice once collected now appear as steep-sided cwms or corries. Huge amounts of rock, stones and soil called moraine, which had been carried by the mass of moving ice, were dumped as the glacier melted at the sides and the end of the valley while smaller material like sand and clay was carried further from the valleys by meltwater flowing in rivers from the ice.

The Ogwr Valleys are the result of thousands of years of the forces of nature at work on the landscape and even today as the rivers flow down the centre of the Valleys, still slowly wearing away the land, the process of natural landscape change continues.

Although this formation process has occurred over millions of years, the Ogwr is in human terms a 'young' valley, being one of the last in Wales to be industrialised and inhabited by large numbers of people.

The first evidence of human presence in the Valleys dates back thousands of years to the stone and bronze ages with weapons such as spearheads being unearthed at Llangeinor and an ancient burial cairn discovered at Bettws. Little is known of the Valleys' early settlers although a monastery, or llan, was founded by Ceinor in the fifth century (which led to the present day name of Llangeinor).

The first written evidence comes from the eleventh century as the Normans, led by Robert Fitzhamon, swept into Wales taking the fertile, hospitable plain of the Vale of Glamorgan and driving the Welsh onto the hilly plateau of Blaenau Morgannwg.

Ownership of the land was claimed by the Norman rulers, but over the centuries was vested in powerful monasteries and wealthy men, while the few tenant settlers scraped a living from the land.

During this time, the upland landscape of the Ogwr Valleys would have been populated by Welsh hill farming communities, with stone buildings and wall boundaries constructed for the farmers and their livestock. Sheep and cattle were the main means of revenue, as wool and milk were central to survival.

Primitive industries were also established as farmers gained extra income for tanning, weaving and cordwaining (shoemaking).

The Ogmore and Rhondda Valleys saw their mineral deposits being extracted before those in the Garw, which was largely left untouched because a poor transport system meant that there was no economical way of carrying coal out of the valley. This finally changed as the coal industry and the development of the Garw Branch railway line grew together when the $5\frac{3}{4}$ mile line from Brynmenyn opened in 1876.

The Ogwr Valleys of Garw, Llynfi and Ogmore have, over the past 150 years, shared an almost identical experience. Essentially rural valleys with a few small hill farms, by the mid-nineteenth century they were engulfed by the 'coal rush' as it swept across South Wales. Industrialisation and mass migration ensued as thousands answered the call of an industry which would change the face of the land as well as the lives of the people. Its legacy is now as important as the changes it brought all those years ago.

The memories and visions of the people of the Valleys at a time of enormous social, economic and cultural change are equal only to that during the exploitation of the South Wales Coalfield, which began less than two centuries ago. It is based on the premise that ordinary people are the only real experts in the histories of their own lives and that it is the apparently insignificant memories – the personal triumphs, the tragedies, the moments of humour and of pain – that tell us what it really means to live and work in the South Wales Valleys.

People poured into the Garw Valley with the sudden development of large-scale industry and the attendant jobs. From a population of 292 people in 1832, the total rose to 8,004 in 1891, with many moving in from all over South Wales and south-west England in the search for work. The changes in the landscape were dramatic as house construction boomed and the towns that exist today were established quickly along the length of the valley. This, coupled with the appearance of the railway, along with the buildings, excavations and waste heaps of the mines themselves, meant that the Garw Valley landscape, which had changed gradually over the years, was suddenly and drastically altered by human activity. For the next forty years the industry thrived and seventeen collieries were established in the Garw Valley. The great rush to produce coal led to poor conditions for health and safety and as mines got deeper in search of new seams of coal, fatal gas explosions became more common.

Alexandra Bowen, Director of Valley and Vale Community Arts, Betws.

Area Managers and Output

In 1961, the Maesteg No.2 area general manager was W.B. Cleaver; the assistant area general manager was R.G. Davies; and the area production manager was D.J. Llewellyn. Group managers were: south, R.E. Petty; Maesteg, C.P. Jones; Garw, V.C. Jones; Ogmore, R.A. Evans. South Group Collieries, Llanharan, Newlands and Werntarw, had a total output of 309,771 tons and a total manpower of 1,634. Maesteg Group Collieries, Bryn, Caerau, Coegnant and St John's, had a total output of 768,562 tons and a total manpower of 2,720. Garw Group Collieries, Ffaldau, Garw, Glengarw and International, had a total output of 446,898 tons and a total manpower of 1,905. Ogmore Group Collieries, Penllwyngwent, Wyndham and Western had a total output of 505,887 tons and a total manpower of 1,993.

Western Colliery, Nantymoel, Ogmore Vale, Glamorganshire, in the 1930s. The Glamorganshire Coalfield is assumed by various authorities to have been approximately 518 square miles. Western Colliery was sunk in 1872 by David Davies & Co., the forerunner of the Ocean Coal Co.

Left: Stables in the Nine-Feet Coal Seam at the Western Colliery in 1967. In 1967 the Western manager was V.W.J. Uppington (6,442 First Class) and the undermanager was J. Dinham (8,967 First Class). *Right*: Wyndham/Western Colliery Underground Double Parting (two sets of dramroads) in April 1979.

The Western Colliery Downcast Shaft was 728ft above Ordnance Datum (OD); it was sited 400yds south east of Nantymoel Church. The coal seams were: Hafod Bottom Coal at 69ft; Abergorki Bottom Coal at 238ft 2in; Pentre Rider at 298ft 1in; Pentre at 339ft 9in; Lower Pentre at 363ft 9in; Caedavid at 456ft 4in; Two-Feet-Nine at 776ft 6in; Upper Six-Feet at 896ft 6in; Lower Six-Feet at 958ft 10in; Caerau at 965ft 4in; Upper Nine-Feet at 1,047ft.

Wyndham Colliery, No.2 Pit, was 618ft OD. The coal seams were: Two-Feet-Nine at 481ft 11in; Six-Feet in overlap at 692ft 5in; Lower Nine-Feet at 891ft 1in; Bute at 938ft. Sunk to 1,076ft 11in, Wyndham Colliery was sunk in 1865 by the Llynfi, Tondu & Ogmore Vale Iron Co., which later became North's Navigation Colliery (1889) Ltd. It was sold to the Cory Bros in 1906. Powell Duffryn Associated Collieries Ltd (PDs, also called the Poverty & Dole group) owned the colliery prior to nationalisation (Vesting Day) on 1 January 1947. It merged with the Western Colliery in 1965. In 1970-1971 the highest output in the South Wales Coalfield at this time was achieved; with a combined manpower of 1,188 Wyndham/Western Colliery produced 467,371 tons. The combined manpower in 1974 was 1,186. In 1981 all coalfaces were in the Five-Feet seam, all coal cutters were ranging drum shearers and all roof supports were powered supports. The Five-Feet seam K3 coalface was 146yds long with a section of 5ft: daily advance 5ft; daily shifts two; daily output 400 tons. The K51 was 218yds long with a section of 4ft 9in; daily advance 3ft 6in; daily shifts two; daily output 500 tons. The K52 was 142yds in length with section of 4ft 9in; daily advance 6ft; daily shifts two to replace K51. Saleable output from development was 30 tons; total colliery saleable output 884 tons per day; output per manshift per coalface 4 tons 8cwt per day; output per manshift overall 1 ton 4cwt; manpower on books 670.

At Wyndham Colliery on 2 March 1868 the accidents reports show that twenty-nine-year-old miner Robert Thomas was killed by a segment; on 13 April 1872 the accidents reports show that twenty-five-year-old collier Jenkin Jenkins was killed by a fall of coal. In 1913 the colliery employed 1,395 miners.

Left: Wyndham Miners in the 1920s. *Right*: Wyndham/Western Machine Power Loading Coalface Conveyor and Roof Supports in 1977. Wyndham Colliery was sunk in 1865 and the Western Colliery was sunk in 1872. Before the passing of the Ogmore Valley Railways Act in 1863, coal for the Tondu Ironworks was carried from the valley on the backs of mules over steep mountain tracks.

The construction of a railway enabled James Brogden to go ahead with the sinking of Wyndham Colliery at the head of the valley. By 1886, when Brogden went into liquidation, the mine was bought by Col. North and became one of the collieries which made up the North's Navigation Co., formed in 1888. At Wyndham Colliery, on 14 April 1886, the accidents reports show that thirty-nine-year-old repairer John Williams was killed by a fall of stone. At Western Colliery on 3 May 1887 the accidents reports show that forty-nine-year-old miner D. Jeremiah was fatally injured.

Both collieries became big producers and big employers and when Wyndham was sold to Cory Bros in 1906, its annual output stood at over 300,000 tons. In 1942 it passed into the Powell Duffryn empire and remained in their ownership until nationalisation. Western remained in Ocean Coal Co.'s hands until nationalisation.

In 1957, the National Coal Board (NCB) approved a massive £3 million scheme to deepen both mines to the 1,440ft 'Mauve Level' and by 1965, they were linked underground to form a single, streamlined unit employing around 900 men. In 1967 the Wyndham manager was T.M. Bond (6,288 First Class) and the undermanagers were W.J. Cook (5,165 First Class) and J.B. Lewis (9,132 First Class).

In 1976 the mining programme worked an area of around sixteen square miles, bounded on the east by the old workings of the Parc, Dare and Cambrian collieries, they were closed in 1966 by the NCB, and on the west by the 210ft Tableland fault. The fault formed a natural boundary with neighbouring Garw/Ffaldau colliery at the head of the Garw Valley. The coal 'take' was split by three major faults; the 19-Yard varying between 18ft 6in and 29ft 6in; the Glyncorrwg, 89 yards; and the Aber, 209 yards.

The production in 1976 was concentrated in the Bute and Gellideg seams at a maximum depth of around 2,000ft, to produce an annual output around 246,050 tons. This was of good quality semi-bituminous coals suitable for blending into coking 'mixes' plus a small percentage for the domestic market.

Wyndham/Western Colliery in 1980. Within the mining operation, there were over eight miles of underground roadways, including one mile over which 100hp locomotives operated and around five miles of high speed belt conveyors, speeding the coal to the shafts and thence to the Ogmore Washery for treatment and grading. At the time, a development scheme, opening up new coalfaces in the Five-Feet seam, was installed at a cost of £1.7 million, with the very latest mining equipment.

In 1976, with a manpower of 1,000, it produced an annual saleable output of 246,050 tons; it produced an average weekly saleable output of 4,413 tons; average output per man/shift at the coalface was 88cwt; average output per man/shift overall was 23cwt; deepest working level was 2,000ft; number of coal faces working was two. Western No.1 shaft had a depth of 1,440ft and a diameter of 15ft. No.2 had a shaft depth of 1,063ft and a diameter of 14ft. Wyndham No.3 had a shaft depth of 1,355ft and a diameter of 15ft. No.4 had a shaft depth of 938ft and a diameter of 12ft. The manwinding capacity per cage wind was 24/28; coal winding capacity per cage wind was 3 tons; winding engines horsepower was 600/1,700; stocking capacity on pit surface was 246,050 tons; average weekly washery throughput was 8,858 tons. The types of coal mined were semi-bituminous and the markets were steel/domestic. It had a fan capacity of 470,000cu.ft per minute; average maximum demand of electrical power was 3,900kW; total capital value of plant and machinery in use was £1.2 million; and estimate workable coal reserves were 12 million tons.

Wyndham/Western Colliery Coal Seams

Depth	Standard Name	Thickness
254yds	Two-Feet-Nine	4ft 5in
314yds	Lower Six-Feet	6ft 5in
345yds	Upper Nine-Feet	6ft
359yds	Lower Nine-Feet	4ft
378yds	Bute	4ft
427yds	Yard	5ft
440yds	Seven-Feet	5ft
460yds	Five-Feet	6ft
477yds	Gellideg	4ft 5in

Wyndham/Western Colliery was closed in 1985 by the NCB.

Penllwyngwent Drift Mine Miners, Ogmore Vale, in 1906. A drift is an entrance tunnel into a mine, which is driven through strata and seams into the required seam, normally at a downward inclination. Penllwyngwent Drift Mine was opened at the beginning of the twentieth century by Cory Bros & Co. Ltd. In 1947 the manpower was 527; in 1955 with a manpower of 449 it produced 114,658 tons and in 1956 with a manpower of 446 it produced 103,975 tons.

Penllwyngwent Drift Mine manriding spake in 1930. A spake is a journey of manriding drams or minecars with seating facilities, used to carry men into and out from their working districts. In 1961 with a manpower of 376 the drift produced 82,726 tons. In 1967 the Penllwyngwent manager was G.A. Hooper (5,814 First Class) and the undermanager was E. Thomas (7,042 Second Class). The seams worked were the Pentre, Two-Feet-Nine and the Yard. Penllwyngwent Drift Mine was closed in February 1969 by the NCB.

Rhondda Main Colliery, Ogmore Vale, in 1911. Rhondda Main Colliery, Catherine Pit, Ogmore Vale, was 385ft OD. Rhondda No.2 was at 749ft, sunk to 817ft 7in. Rhondda Main Colliery was sunk by Lewis Merthyr Colliery Co. in 1909. The colliery was plagued by serious water problems causing the colliery to close in 1924 by the Lewis Merthyr Colliery Co.

Aber Colliery, Tynewydd, Ogmore Vale, in 1910. Aber Colliery was the most southerly of the quartet of Cory Bros' mines in the Ogmore Valley. The others were Cwmfuwch, Penllwyngwent and the Wyndham. The colliery was owned by the Aber & Ynysawdre Coal & Coke Co. Ltd and the Ocean Coal Co. At approximately 1.00 a.m. on Monday 14 May 1888, the accidents reports show that five miners were killed and two injured by an explosion. A horse was also killed. Aber Colliery opened in 1867 and was closed when acquired on 1 January 1947 by the NCB.

International Colliery, Blaengarw, Garw Valley, in 1937. International Colliery No.1 Downcast Shaft was 737ft OD. Deepening to 1,436ft 9in. In 1967 the International manager was E. Morgan (5,122 First Class) and the undermanager was D.W. Williams (6,270 First Class).

Coalcutting in the Victoria Seam in 1937. No.2 Upcast Shaft was 739ft OD. International Colliery was opened in 1883 by the International Coal Co. Ltd. In 1928 it was owned by Glenavon Garw Collieries Ltd, in 1937, the Ocean Coal Co., Ocean & United National Co. and the NCB. The colliery consisted of two shafts, a level, with a third shaft being sunk in 1910; the level closed in 1923. It was later linked with the Garw Colliery. In 1889 the colliery employed 916 miners. International Colliery was closed on 4 November 1967 by the NCB.

Garw Ocean Colliery, Blaengarw, Garw Valley, in 1906. Garw Colliery Downcast Shaft was 699ft OD. Seams worked were: Victoria at 88ft 7in; Upper Yard at 152ft 10in; Caedavid at 248ft 5in; Two-Feet-Nine at 551ft 4in; Lower Six-Feet at 714ft 1in; Caerau at 728ft 1in; Upper Nine-Feet at 813ft 5in; Lower Nine-Feet at 898ft 5in; Bute 946ft 11in; Yard at 1,071ft 11in; Upper Five-Feet at 1,157ft 2in; Lower Five-Feet at 1,179ft 5in, sunk to 1,201ft 10in. Garw Colliery Upcast Shaft was 700ft OD and it was sited 650yds north east of Mount Zion Church, Blaengarw. Seams worked were: Two-Feet-Nine at 566ft 1in; Lower Six-Feet at 699ft 1in; Upper Nine-Feet at 802ft 3in; Lower Nine-Feet at 875ft 6in; Yard at 1,098ft 3in, sunk to 1,170ft 10in.

Garw Ocean Colliery was sunk in 1884 by the Blaengarw Ocean Coal Co., a subsidiary of the Ocean Coal Co., and followed that company into its merger with United National Collieries. The accidents reports show that on 19 February 1884, thirty-four-year-old sinker Richard Jones was killed by a fall from the side of the pit; on 7 June 1886, the accidents reports show that twenty-two-year-old collier William Jones was killed by a fall of roof; and on 17 May 1887 the accidents reports also show that twenty-six-year-old rippers William Roberts and David Griffith were fatally injured.

In 1967 the Garw manager was T. Clarke (5,299 First Class) and the undermanager was D.N. Thomas (8,384 First Class). Under NCB control it was merged with Ffaldau Colliery in 1975. The districts worked with Ffaldau Colliery were the Lower Nine-Feet seam M3 coalface 190yds in length and was stopped coaling by a fault. The M4 was 191yds in length with a section of 5ft; daily advance 5ft; daily shifts two; daily output 300 tons. The M5 was 240yds in length with a section of 4ft 3in. The Bute seam B1 was 218yds in length with section of 4ft 3in; daily advance 5ft 2in; daily shifts two; daily output 500 tons. Gellideg seam G73 was 218yds in length with a section of 4ft 9in; daily advance 6ft 6in, stopped. G72 was 125yds in length with a section of 5ft; daily advance 6ft 6in; daily shifts two; daily output 450 tons. G70 was 218yds in length with a section of 5ft; daily advance 1.25m; daily shifts two; daily output 450 tons. B10 had a continuous mining cutter (JCM) with a daily output of 75 tons. Output per manshift per coalface 5 tons 3cwt; output per manshift overall 1 ton 7cwt; manshift per day per development 176; manshift per day per coalface 230; elsewhere below ground 185; manshift per day per surface 138; total 729. All coalfaces had powered supports and coal cutting was carried out by two Ranging Drum Shearers. Garw Ocean Colliery and Ffaldau Colliery was closed in 1985 by the NCB.

Ffaldau Colliery, Pontycymer, Garw Valley, in 1910. Ffaldau Colliery, also known as the Oriental, Downcast Shaft was 519ft OD. Seams worked were: Upper Yard Pentre at 52ft 3in; Two-And-A-Half at 98ft 3in; Eighteen-Inch at 112ft 10in; Caedavid at 149ft; Two-Feet-Nine at 419ft 2in; Upper Six-Feet at 570ft 11in; Lower Six-Feet at 645ft 1in, sunk to 663ft 7in. Ffaldau Colliery Upcast Shaft was 518ft OD and sited 550yds north west of St Theodore's Church, Pontycymer. Seams worked were Caedavid at 117ft 7in; Two-Feet-Nine at 407ft 11in; Upper Six-Feet at 549ft 10in; Lower Six-Feet at 638ft 4in; Caerau at 646ft, Lower Nine-Feet at 721ft 3in; Bute at 758ft 9in; Yard at 863ft 11in; Middle Seven-Feet at 925ft 10in; Lower Seven-Feet at 954ft 11in; Upper Five-Feet at 973ft 7in, sunk to 995ft.

Ffaldau Colliery, Victoria or Braich-Y-Cymmer Shaft was 528ft OD. Seams worked were: Caedavid at 173ft 10in; Two-Feet-Nine at 534ft 2n; Upper Six-Feet at 661ft 2in; Lower Six-Feet at 752ft 3in; Caerau at 759ft 1in; Lower Nine-Feet at 820ft 4in; Bute at 877ft 9in; Yard at 974ft 5in; Middle Seven-Feet at 1,033ft 1in; Lower Seven-Feet at 1,068ft 4in; Five-Feet at 1,091ft 5in, sunk to 1,104ft 11in.

Ffaldau Colliery was opened by the Ffaldau Colliery Co. in 1876 and owned by Cory Bros & Co. Ltd prior to nationalisation. On 12 May 1884 the accident reports show that forty-seven-year-old collier Thomas Griffiths was killed by a fall of stone; on 18 June 1884 they show that nineteen-year-old colliery Richard Price was killed by a fall of stone; on 24 May 1887 thirty-five-year-old collier Thomas Powell was killed; and on 5 January 1888 fourteen-year-old collier Thomas Lewis was also killed. In conjunction with the Braich-Y-Cymmer Colliery the seams worked in 1947 were the Yard, Nine-Feet, Gellideg and Seven-Feet and the manpower was 395. In 1954 with a manpower of 692 it produced 269,154 tons; in 1955 with a manpower of 775 it produced 258,467 tons; in 1956 with a manpower of 793 it produced 262,963 tons; in 1957 with a manpower of 800 it produced 274,836 tons; in 1958 with a manpower of 779 it produced 244,641 tons; in 1960 with a manpower of 887 it produced 240,813 tons and in 1961 with a manpower of 950 it produced 277,844 tons. In 1967 the Ffaldau manager was N. Burton (6,981 First Class) and the undermanager was S.D.V. Williams (9,220 First Class). The NCB merged Ffaldau with Garw Colliery in April 1975. The seams also worked with Ffaldau Colliery were: the Yard; Nine-Feet; Gellideg; and Seven-Feet.

19

Ffaldau Colliery Rescue Team in the 1920s. On 20 December 1985, over 2,000 people marched through the Garw Valley on the last day of operation of the Garw/Ffaldau Colliery in Blaengarw. This day marked the end of 100 years of coalmining in the Valley.

Anderson Boyes Coalcutting Machine in 1962. The emotional day was organised by the local NUM branch to show that the comradeship and loyalty evident in the 1984/85 strike was irrepressible and would be a major factor in the rebuilding and on-going life in the valley. Ffaldau Colliery was closed on 20 December 1985 by the NCB.

Braich-Y-Cymmer Colliery, Pontycymer, Garw Valley, in 1947. Braich-Y-Cymmer Colliery was sunk in 1890 by the Ffaldau Colliery Co. and worked together with the Ffaldau Pits, which merged with the Garw Colliery in 1975. Seams worked were the Caedavid; Two-Feet-Nine; Upper Six-Feet; Lower Six-Feet; Caerau; Lower Nine-Feet; Bute; Yard; Middle Seven-Feet; Lower Seven-Feet; Five-Feet, sunk to 1,104ft 11in. Braich-Y-Cymmer Colliery was closed in 1985 by the NCB.

Glengarw Colliery, Blaengarw, Garw Valley, in 1912. Glengarw Colliery was worked by the Glenavon Garw Collieries Ltd and the NCB. In 1913 the colliery employed 391 miners; in 1954 with a manpower of 398 it produced 73,000 tons; in 1957 with a manpower of 305 it produced 56,813 tons; and in 1958 with a manpower of 294 it produced 56,721 tons. The coal seams worked were the Victoria, Yard and the Caedavid. Glengarw Colliery was closed in November 1959 by the NCB.

Darren Colliery Tip, Blaengarw, Garw Valley, in the 1930s. When the colliery closed men and boys turned it into the playing field it now is. In 1888 the Darren manager was David John (1,215 First Class), undermanager John Griffiths (936 Second Class), undermanager William Howells (937 Second Class). The colliery worked the Rhondda No.3 seam with a manpower of 143 and the Victoria seam with a manpower of 107. Darren Colliery was opened in 1888 and was closed when acquired on Vesting Day 1947 by the NCB.

Duchy No.3 Level, Pontyrhyl, Garw Valley, in 1928. Duchy No.3 Level was opened in 1924 by the Duchy Colliery Co. Life underground, down in the bowels of the earth, was hazardous but coalmining was a daily and regular feature for me all my working life. If asked to describe life underground I would say, 'It's not easy with the appalling conditions, roof breaking, timber creaking, stones falling, coal and stone dust rising, water dripping, poor ventilation, using naked flame lamps and inadequate safety flame lamps giving insufficient light (about one candle light), sweat running, blood seething and, on some occasions, breath failing.' Yet this was the way we earned our bread and butter. Duchy No.3 Level was closed when acquired on Vesting Day 1947 by the NCB.

Waun Bant Level, Garw Valley, in 1897. A level is a tunnel driven horizontally or on a slight gradient to connect underground workings with the surface. The coal was then mined without having to dig a shaft and is very common in hilly areas. The Peg and Ball naked flame lights were used in this mine. No females were legally allowed to work underground when Lord Shaftesbury's Coal Mines Act of 1839 was passed, but there were times when the rules were ignored and many years were to pass before complete success was to be achieved. An Investigating Committee in 1842 still found instances of children aged four, five and six employed underground. Waun Bant Level was abandoned in 1933.

Repairers at work in difficult roof conditions in the 1930s. The number of hours worked underground was twelve hours per day at the end of the nineteenth century, this was then reduced to ten hours per day. The objective of the workforce was for a reduction to eight hours per day. A jingle was written: 'Eight hours work, Eight hours play, Eight hours sleep, And eight bob a day.' Records show that hundreds of young coalminers had to cease work due to the dreaded disease of silicosis, which was considered a progressive disease due to the lungs hardening like cement. It caused painful suffering and often culminated in loss of life, very often before the sufferer had reached his twentieth birthday, indeed a very young age: '...and they worked us to death...'

Caerau Colliery, Caerau, Lynfi Valley in 1930. Seams worked were: Upper Blackband Ironstone at 148ft 9in; Victoria at 306ft 10in; Upper Yard at 387ft 10in; Two-And-A-Half at 423ft 10in; Caedavid at 481ft; Two-Feet-Nine at 813ft 7in; Four-Feet at 898ft 7in, sunk through small fault to Upper Six-Feet at 971ft 1in; Lower Six-Feet at 1,020ft 6in; Caerau Vein at 1,051ft 6in, sunk to 1,066ft 6in.

Caerau Colliery, No.2 South Pit was 769ft OD. Seams worked were: Victoria coal seam at 235ft 2in; Upper Yard at 328ft 8in; Two-And-A-Half at 360ft 4in; Caedavid Bottom Coal at 419ft 9in, sunk through a small fault to Two-Feet-Nine at 753ft; Four-Feet at 823ft 9in; Upper Six-Feet at 923ft 9in; Lower Six-Feet at 938ft 11in; Caerau Vein at 1,015ft 6in, sunk to 1,052ft 7in.

Caerau Colliery, No.3 Pit was sunk in 1889. Seams worked were: Clay at 56ft 7in; Victoria coal seam at 322ft 10in; Two-And-A-Half at 436ft 11in; Caedavid Bottom Coal at 499ft 9in, sunk to 517ft.

Caerau Colliery was opened in 1891 by North's Navigation Collieries Ltd. On 7 January 1890 the accident reports show that forty-seven-year-old sinker Thomas Jones was killed by a banking trolley falling down the shaft; they also show that fifty-one-year-old surface labourer Michael Constantine died on 18 December by falling off a 5ft wall; and on 20 April 1891 they also show that thirty-four-year-old surface worker David Williams was killed when a pump was being lifted by a crane from a truck when a link of the chain broke and the pump struck him when falling.

In 1913 the colliery employed 1,998 miners; in 1947 the manpower was 638; in 1954 with a manpower of 806 it produced 167,290 tons; in 1955 with a manpower of 827 it produced 159,688 tons; in 1956 with a manpower of 769 it produced 150,182 tons; in 1957 with a manpower of 765 it produced 140,499 tons; in 1958 with a manpower of 732 it produced 173,166 tons and in 1961 with a manpower of 641 it produced 122,244 tons. In 1967 the Caerau manager was L.S Rees (3,648 First Class) and the undermanager was W.A. Hurd (5,686 First Class). The coal seams worked were the Red Vein, Lower New, Upper New and Harvey. Caerau Colliery was closed on 26 July 1977 by the NCB.

Coegnant Colliery, near Caerau, in 1980. Following the publication in 1973 of Plan for Coal, a massive capital investment programme was launched by the NCB. This colliery was one of those involved in the programme. Work on sinking the twin shafts of Coegnant Colliery began more than 100 years ago in 1882 to the Lower New seam at a depth of 244yds by Llynfi Coal & Iron Co. Ltd, and later North's Navigation Collieries Ltd, to work the rich Six-Feet seam, lying 1,200ft below the Llynfi Valley. The first coal emerged from Coegnant in 1890. In 1976 Coegnant was producing an annual average output of around 100,000 tons of semi-coking coal from the Lower Nine-Feet seam. This was taken to the Maesteg central washery, where it was blended into coking 'mixtures' for supply to the steel industry. Originally, Coegnant shared its 'take' area with neighbouring Caerau and St John's Collieries, but the operations in 1976 moved west across the extensive Penycastell Fault, which lies along the bottom of the Llynfi Valley and divides the coal deposits, causing drops in the same coal seam of between 152ft and 600ft. An odd geological accident in one area resulted in the break bringing the Six-Feet exactly opposite the Five-Feet, two seams that, in other places, are around 594ft apart. Faulting was a constant problem in South Wales mining, caused 100 men to leave Coegnant in 1976, to work at nearby St John's, whilst 1,200m of underground roadways were driven in search of additional reserves in the Lower Nine-Feet seam. There were more than four miles of underground roadways within the mining programme, carrying around a mile and a half of high-speed belt conveyors. The average length of a coalface was 153yds and each face carried more than £200,000 worth of sophisticated mining equipment. Coegnant, in fact, was amongst the first South Wales mines to pioneer the technique of two shearers on the same coalface, as long ago as 1969. Among considerations in 1976 for the future, was a project to exploit reserves in the almost-virgin Yard seam to the east of the shafts.

In 1976, with a manpower of 690, it produced an annual saleable output of 984,200 tons; it produced a weekly saleable output of 2,066 tons; average output per man/shift at the coalface was 2 tons 3cwt; average output per man/shift overall was 1 ton; deepest working level 1,870ft; number of coal faces currently working – two. No.1 shaft: depth 1,244ft; diameter 15ft. No.2 shaft: depth 1,133ft; diameter 20ft; man-winding capacity per cage wind 25; coal-winding capacity per cage wind 3 tons; total capital value of plant and machinery in use £580,000; estimate workable coal reserves 2.5 million tons. Coegnant Colliery was closed on 27 November 1981 by the NCB.

Oakwood Colliery, Llynfi Valley in 1919. Oakwood Colliery was 402ft OD. It was sunk to Bute at 900ft. Oakwood Colliery was opened in 1864 by W. Davis and then owned by: Elders Navigation Collieries Ltd from 1899, Celtic Collieries Ltd from 1909 and North's Navigation Collieries (1889) Ltd, from 1920 until closure in 1929. It was also known as Pwll Davis and employed 250 miners in 1922. On 14 February 1872 fourteen miners were burnt and suffocated by an explosion; naked flame lights were in use at the time. Four horses and two donkeys were also killed. Oakwood Colliery was closed in 1929 by North's Navigation Collieries Ltd.

Bryn Colliery, Bryn, Near Maesteg, in 1920. Bryn Colliery was 468ft OD. Bryn Colliery was opened in 1904 by Baldwins Ltd. The coal seams worked were: Two-Feet-Nine seam at 113ft; Four-Feet at 149ft 11in; Six-Feet at 315ft 1in; sunk through disturbed ground including Jubilee Slide to Lower Seven-Feet at 1,001ft 10in; Upper Five-Feet at 1,026ft 11in; Lower Five-Feet at 1,141ft 10in, sunk to 1,144ft 11in. Bryn Colliery was closed on 7 June 1963 by the NCB.

Left: St John's Colliery, Maesteg, in 1954. *Right*: St John's Colliery saw mill, 20 July 1977. The coal seams worked were: Two-Feet-Nine at 457ft 9in; Upper Four-Feet at 495ft 9in; Lower Four-Feet at 561ft; Upper Six-Feet at 656ft 2in; Lower Six-Feet at 716ft 1in; Caerau at 742ft 7in; sunk through disturbed ground including Jubilee Slide to Lower Seven-Feet at 1,088ft 3in; Upper Five-feet at 1,117ft 9in, sunk to 1,135ft 9in. North Pit was 638ft OD and the site was 2,460yds E 7° of Maesteg Railway Station. National Grid Ref. 87589171. Two-Feet-Nine at 495ft 11in; Upper Four-Feet at 532ft 5in; Lower Four-Feet at 623ft 7in; Upper Six-Feet at 701ft 9in; Six-Feet at 759ft 6in; sunk through disturbed ground including Jubilee Slide to Lower Seven-Feet at 1,088ft 3in; Upper Five-Feet at 1,159ft 5in, sunk to 1,165ft.

St John's Colliery was sunk in 1865. From 1920 it was owned by North's Navigation Collieries Ltd. In 1940 the colliery employed 1,480 miners; in 1956 with a manpower of 985 it produced 246,518 tons; in 1958 with a manpower of 970 it produced 333,635 tons; and in 1961 with a manpower of 843 it produced 325,281 tons. In 1967 the St John's manager was A.J. Reed (6,221 First Class) and the undermanager was R. Thomas (6,058 Second Class). In 1981 all coal cutters were Ranging Drum Shearers; all roof supports were powered supports with the exception of M16, which had posts and bars. The Lower Six-Feet seam S11 coalface was 120yds in length with a section of 4ft 3in; the daily advance was 4ft 6in; daily shifts were two; daily output was 400 tons. The Lower Nine-Feet seam M16 coalface was 150yds in length with a section of 5ft; the daily advance was 5ft 2in; daily shifts were two; daily output was 450 tons. The M17 was 142yds in length with a section of 5ft; it had a daily advance of 6ft; daily shifts were two; daily output was 470 tons. The M18 was 87yds in length with a section of 5ft and was a standby face. The Bute seam B1 was 240yds in length with section of 4ft 1in; daily advance was 5ft. Gellideg seam G38 was 207yds in length with a section of 4ft 9in; daily advance was 5ft 6in; daily shifts were 3/2; daily output was 460 tons. G40 was 207yds in length with a section of 4ft 9in; daily advance was 5ft; daily shifts were two; daily output was 460 tons from January 1982. Saleable output from development was 30 tons per day; total colliery output from May to August was 1,252 tons; output per manshift per coalface was 4 tons 2cwt; output per manshift per overall was 1 ton 8cwt. St John's Colliery, the last deep mine to work in the Llynfi Valley, was closed on 22 November 1985 by the NCB.

Garth Colliery, near Maesteg, in 1906. Garth Colliery was opened in 1865 by J. Brogden & Sons and later worked by Celtic Collieries Ltd and North's Navigation Collieries Ltd. When the colliery was owned by the Celtic Collieries Co. it was known as Garth Celtic. It was deepened to the lower seams in 1882 and at its peak had sixty coke ovens and employed 616 men. Garth Colliery was closed in 1930 by North's Navigation Collieries Ltd.

Left: Garth Colliery Gaffers (Officials) in the 1920s. Official was the generic term for all levels of management, from agent down to shotfirer (shotsman). Formerly, in large coal companies one or more agents would have been in charge of a group of mines. Each mine would have had a manager, who was required to be properly qualified and answerable to the Inspector of Mines. The undermanager (overviewer) was generally responsible for the underground activities at the mine and, like the managers, was legally required to be properly qualified in respect of his supervision of operations in his district. The overman was responsible for the provision of supplies including timber for supporting the roof; at one time he also had responsibility for calculating the wages due to each collier and all other miners in his district. The deputy (fireman) was a qualified official and was legally in charge of, and responsible for, his district and all its activities, including production, safety, supplies, ventilation, etc., under the regulations and the Mines and Quarries Acts. The shotsman was a qualified official who fired shot holes in a district.

Cwmfelin Boot Repairing Depot, near Maesteg, for miners' children during the 1926 strike. Fellow workers decided to support the miners and a General Strike began on 3 May 1926, which threatened to bring the whole of Britain to a standstill. A sense of alarm spread through that section of society that supported the Government and the coalowners. They felt the country stood on the brink of a revolution. Boot centres were established which repaired the shoes of miners and their children.

Cefn-Y-Bryn Colliery, near Maesteg, in 1920. The miners are wearing the Peg and Ball naked flame light on their caps; there were no safety helmets worn at this mine to protect the miners from head injuries. In 1960 at Tylorstown No.9 Colliery, Rhondda Fach, several miners and colliers at the coalface where I worked also wore this type of cap. Cefn-Y-Bryn Colliery was worked by the Cefn-Y-Bryn Colliery Co. at the beginning of the twentieth century. Cefn-Y-Bryn Colliery was abandoned on 5 December 1924.

Coytrahen Level in 1965. The method of working at this level was by pillar and stall. A main heading was driven from the level into the seam. Then at intervals, along one or both sides of this main heading, smaller headings were turned off at right angles. From these cross headings, stalls were cut at intervals and it was by the excavation of these stalls that the bulk of the coal was won. As the seam was penetrated, square, or almost square, pillars of coal were left standing about every 10yds in order to support the roof. In size, these pillars depended on the depth of the working, the hardness of the seam and the strength of the roof and floor. Coytrahen Level was abandoned in 1928.

Left: Inspectors (HMI) and Mines Rescue Teams alkaline battery-powered Safety Hand Lamp in the 1930s. *Right*: Pentwyn Level, Coytrahen on the Llynfi tramroad in 1965. Pentwyn Level was abandoned in 1928. In the early days of mining the transportation of coal proved a formidable task. The use of two-wheeled carts was very limited owing to the ruggedness of the terrain. Packhorses carried the coal from where it was produced to more centralised collection points.

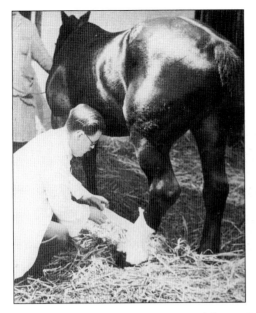 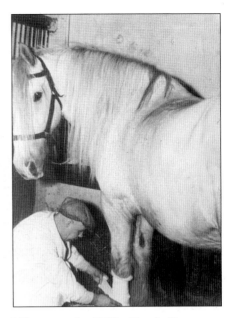

Left and Right: Tondu Horse Hospital for Pit Ponies and Horses in the 1950s. Tondu House was built in the early 1600s and contained longstanding veterinary facilities in the outbuildings. Tondu House was demolished in 1963. In 1930 there were approximately 11,500 horses employed underground in the South Wales Coalfield. Several pit ponies had over eight years service underground, were treated like family pets by the hauliers and, not to be left out, some horses enjoyed a chew of tobacco to keep the dust at bay.

Tondu Horse Hospital for Pit Ponies and Horses in the 1950s. Pit ponies were well cared for by the hauliers; stables generally were warm, dry and comfortable and many were lit by electricity. Moss litter or sawdust was always provided for bedding. Plenty of good food and clean water was at hand, both at the stables and while at work, and a sufficient supply of medicine and dressing was readily available.

Felin Fach Colliery Waterwheel, Tondu in 1918. The wheel was used for pumping mine water from the colliery in 1859. On 21 August 1860 the accidents reports show that thirty-eight-year-old collier David Jones was killed by a fall of coal from the side of his working place. In 1870 the South Wales Coalfield production exceeded 13,590,000 tons, fifty per cent for the export trade, and mining by the longwall method replaced the more traditional pillar and stall technique. Felin Fach Colliery was abandoned in 1931.

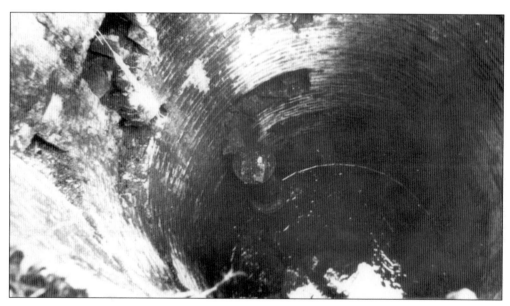

Felin Fach Colliery Shaft in 1958. The coal seam worked extensively by the colliery was the Cribbwr Fach. The Cribbwr Fach seam was formerly worked from Bell Pits near its crop in the vicinity of the Porthcawl Branch Railway between Kenfig Hill and Cefn Junction. Further east several drift mines worked the coal on a small scale, as for example at Bankershill Slant, which was 1,070yds W 19° S of Parc Slip Colliery. The seam is said to vary from 2ft 3in to 3ft of clean coal. Bell Pits were worked in various areas of the South Wales Coalfield.

Brynmenin Collieries Rescue Team, Brynmenyn, in 1920. In 1886 the Royal Commission recommended the establishment of mines rescue stations. They did not become widespread until the Coal Mines Act of 1911 made them compulsory. In 1877 in the South Wales Coalfield a total of 159 men and boys lost their lives, the youngest just twelve years old.

Brynmenin Drift Mine, Brynmenyn, in 1906. Brynmenin Drift Mine was opened in 1901. The colliery was owned by Soloman Andrews & Sons (who also owned and worked it in conjunction with nearby Brynmenin brickworks). The colliery part of the business was eventually closed at a loss to the shareholders of over £46,000. Brynmenin Drift Mine was closed in 1908 by Soloman Andrews & Sons.

Left: Bryncethin Barrow Pit Miners Lewis John and Chas in the 1930s. Bryncethin Barrow Colliery No.3 Pit was 261ft OD. *Right*: Remains of Bryncethin Colliery Barrow Pit in 1950. On 26 September 1874 five sinkers were killed when a suspended shaft staging gave way during sinking operations. Those killed were: S. Llewellyn, aged thirty-six; C. Howell aged twenty-seven; W. Franks, T. Davies and J. Rook, all aged twenty-three. A part of the cast-iron crab-winch broke, the stage canted and seven men fell. Two were saved. At the inquest it was recommended wooden crabs to be substituted for cast-iron ones.

Taking a bath at Bryncethin in 1955. The coal fire has always been a symbol of warm welcome in the Valleys. The fire was also used for baking bread and cakes, cooking dinners, boiling kettles, boiling buckets of water for the bath and drying clothes. Before the pithead baths were opened, the miner came home from the pit and bathed in a tin bath in front of the fire. Bryncethin Colliery was sunk in 1874. At Bryncethin Colliery on 15 November 1877 a miner was killed by an exploding canister of gunpowder.

Werntarw Colliery, near Llanharan, East of Ogmore Valley, in 1950. Believed to be 900ft deep, Werntarw Colliery was opened in 1915. It was worked by Meiros Collieries Ltd in the 1920s and the South Wales Coalite Co. Ltd prior to nationalisation. The coal seams worked were the Six-Feet, Pentre and Hafod.

Left: Werntarw Colliery Boring Rig in 1967. Exploratory holes were bored to define the coal seams, mine water, methane gas, etc., in the vicinity. Werntarw Colliery was closed in August 1964 by the NCB. *Right*: Bryn Chwith Colliery, Near Bryncethin Man Riding Spake in 1940. Bryn Chwith Level was opened in 1877 by Hedleys Collieries Ltd and later worked by the Raglan Colliery Co. In 1913 the colliery employed ninety-one miners. Bryn Chwith Colliery was closed when acquired on Vesting Day 1947 by the NCB.

Parc Slip Colliery, Fountain, Aberkenfig, 27 August 1992 – the day following the tragic explosion. The disaster occurred at approximately 8:20 a.m. on Friday 26 August 1892; 112 men and boys died in the explosion. It was, without doubt, one of the most important as well as tragic events in the history of the locality. Of the thirty-nine others who 'came out alive' some bore both physical and emotional scars. The victims left behind them over 200 dependants, widows, children and elderly parents.

In August 1892, after several years in the ownership of North's Navigation, Friday 26 August promised, in the anticipation of many, to be a day of pleasure, for it was the day of the annual St Mary Hill Fair. The fair was only too often spoiled by bad weather, but this time the previous couple of days had been fine and Friday, too, dawned bright with promise. Some colliers had changed shifts to go to the fair and altogether there was a good deal of pleasant bustling about to be done as excited families sat down to breakfast or prepared for the day's outing.

The fun of the fair came to an abrupt halt as the thunderous roar of the massive explosion rent the air. It was the last week in August and the people of a small mining community had been enjoying a day out. But as the ground shook beneath them from the blast four miles distant, the hearts of the holidaying crowds were suddenly filled with terror. Miner Ned Jones knew what the explosion meant. It meant his twelve-year-old son was in grave danger. David Jones of Cae Ceddau, Litchard Hill, Bridgend, had been due to go to St Mary Hill Fair with his father that sunny late-summer day. But, in a fateful decision, he opted to go to work at the pit instead. As he heard the sound of the explosion, Ned leapt into the saddle of his horse and galloped to the Parc Slip Colliery where he and his son worked. It was, already, too late. David was one of the 112 men and boys killed in the disaster. His nephew, Wyndham Parry John, of Llangeinor said, 'He was supposed to go with his dad to the fair, but changed his mind.' A great friend of his father's told him not to worry, saying: 'I will look after him, Ned,' 'Don't worry, Ned, he's safe in my arms.' When David was found, his father's friend had been as good as his word. His arms were wrapped around him trying to shield him from the explosion. 'The explosion was heard by Uncle Ned at St Mary Hill. He knew it was Parc Slip. He mounted his horse and galloped to Parc Slip. The horse dropped dead as he dismounted at the colliery.' Ned Jones struggled to come to terms with the guilt associated with his son's death. 'He used to tell my mam, "It was my fault, Sarah, I should have made him come with me"', said Wyndham John.

For a period of about ten days the whole neighbourhood must have been steeped in communal gloom as the dead bodies were gradually and laboriously retrieved, as multiple funerals hurried through the sombre streets towards one or other of the burial grounds, St John's, Llansantffraid, Smyrna, Nebo, Siloam, where graves had been prepared, sometimes row upon row. The first payments were made out of the relief funds; some families needed assistance even with burial expenses. The domestic devastation is hard as well as painful to imagine. More than one father had died with his son or sons; two or three or even four brothers in one family had met their deaths together. One woman lost father, husband and brother. No aspect of social life was left untouched. Schools were suspended as pupils stayed away; churches and chapels found it all but impossible to continue with services where attempts at singing were choked by tears.

It was two months after the disaster before the last bodies were retrieved, the last victim (George Dunster) being laid to rest in St John's Churchyard on 2 November. The repercussions of the tragedy must, however, be measured in years rather than months. For nearly sixty years afterwards, payments continued to be made out of the Relief Fund, albeit to a steadily diminishing band of dependants. The fund was wound up in April 1950, when the names of only two dependants remained on the books.

Parc Slip was in many ways the least typical of the major pit disasters of the late nineteenth century. The mine was a relatively small one, employing 200 men, and previously had a good safety record in an area not noted for fiery or gassy pits. It was also a drift mine, in which a tunnel led into a hillside, descending gently to the coalface. Most of the danger of the coal industry has been associated with deep mining. Parc Slip, which left sixty women widowed and 153 children fatherless, also became notable for the number of people who got out of it alive. Because the explosion had occurred deep in the bowels of the pit, it was thought that no one working at the coalface would survive. But a series of dramatic rescues succeeded in getting aid and fresh air to the men who had huddled in the darkness for days rather than venture out through the poison-filled tunnels. A total of forty-one men were recovered alive from the pit, although some of the rescuers, like James Bowen, lost their own lives trying to bring them to safety. James Bowen was one of the true heroes of the Parc Slip disaster. There is little doubt that, but for his selfless actions and strong leadership in the wake of the explosion, many more men would have died there. But his brave attempts to get help to his colleagues cost Mr Bowen his own life. Two of his sons also perished in the blast; a third walked out of the pit alive, however, after nearly a week underground. Mr Bowen was in charge of a gang of twenty-two men at the coalface at the number four stage of the pit. He attempted to lead the nineteen who survived the immediate aftermath of the blast to safety. They reached an area of the mine known as the East Turn, but here Mr Bowen stopped. He explained to the rest of the men that the safest course of action would be to stay where they were while he went to get help. 'Some of his companions argued against this, while others offered to go with him but he remained firm, and being the man of authority they obeyed,' recalled an eyewitness in a contemporary newspaper report of the tragedy. Mr Bowen left the gang of men about midday on Friday. Another rescue party discovered his lifeless body later that night. Mr Bowen's sons Jason, sixteen, and Thomas, fourteen, who also died are buried with their father in Penyfai Churchyard. But a third son, Levi, was one of the miners who survived the blast, returning as if from the grave. 'They had given him up for dead but he walked out a week later,' said granddaughter Mrs Walters. 'When he knocked on the door, his mother thought she had seen a ghost.' The Parc Slip disaster of 1892 was the worst to hit the small close-knit mining communities and still looms large in the history of the area. Parc Slip Drift Mine was opened by the Llynfi, Tondu & Ogmore Coal & Iron Co. Ltd in 1875 and was closed in 1904 by North's Navigation Collieries Ltd.

We continue on with our journey by returning to where the story ended at Llantrisant Colliery at the end of Chapter Three in *South Wales Collieries Volume One*. *Left*: Dehewydd Colliery, Llantwit Fardre, in 1900. It was abandoned in 1914. *Right*: Llest Llantwit Colliery, Llantwit Fardre, in 1901. It was abandoned in 1902.

Left: Bryn Colliery, Llantwit Fardre, in 1905. The colliery shaft was only 38yds deep and formerly a water balance pit. Each cage was fitted with a tank, which could be filled with water when it was at the bank (pit top). When it was necessary to raise a dram of coal to the surface, the dram was placed in the cage at pit bottom and the tank of the topmost cage was filled until it was heavy enough to counter-balance the weight of the loaded dram at the pit bottom and raise it to bank. The water in the descending cage was let out at the pit bottom and had to be pumped back to the surface. Bryn Colliery was abandoned in the 1920s. *Right*: Workmen and Gaffers in the 1920s.

Llanharan Colliery, East of Ogmore Valley, in 1950, in the southern boundary (south crop) of the South Wales Coalfield. The shafts were partly sunk in 1873 and were abandoned through lack of funds. It was owned and finally sunk by Powell Duffryn Associated Collieries Ltd in 1922. In 1961, with a manpower of 571, it produced 94,654 tons.

Llanharan Colliery Staple Shaft in 1950. A staple shaft is an underground shaft that connects from one seam to another seam below. Usually vertical, it can be used to transport coal by chute, to all existing transporting systems in the lower seam. Llanharan Colliery was closed in August 1962 by the NCB.

Meiros Colliery, Llanharan, in 1925. Meiros Colliery was 397ft OD. On 11 November 1891 the accidents reports show that forty-year-old collier Thomas Davies, thirty-year-old collier John Parsons and fifteen-year-old collier's boy Dan Lewis were killed by an explosion of firedamp (methane gas). Meiros Colliery was worked in the 1920s by G. & I. Coal Collieries Ltd, which was taken over by the Meiros Collieries Ltd. In 1913 the colliery employed 300 miners.

Left: Meiros Colliery in 1938. The collier is wearing an acetylene carbide lamp, a naked flame light, which was the cause of many explosions. Firedamp is the gas the deputy tests for with a safety lamp. A safety lamp is a lamp in which dangerous contact between the external atmosphere and a naked flame is prevented. Sir Humphrey Davy invented the Davy safety lamp in 1815. His gauze principle was the forerunner of the modern safety lamp. The deputy also uses a probe similar to a walking stick; this is used for taking an air sample from the roof cavities, with the aid of a rubber injector bulb. The sample is injected into the official's lamp. He then estimates the percentage of methane present in the cavity from the height of the gas 'cap' above the flame. Firedamp will burn if ignited and may cause an explosion. Meiros Colliery was abandoned in November 1914.

Right: Llanharry Mine in 1970. This mine produced iron ore; coal was not produced from this pit. The headframe of this iron ore mine is typical of the colliery headframes and the most readily identifiable symbol of the coal industry was its distinctive outline. Its strength, size and material of construction reflected both the ever-increasing depth of shafts and power of the adjacent winding engines. The great depth of some mine shafts and the height of the cages meant that very strong headframes had to be constructed over the shafts. In the early years of the twentieth century a majority of colliery headframes were made of pine. The first wrought-iron lattice headframes were built in the mid-nineteenth century and many of the colliery companies had their own distinctive style.

Llanharry Iron Ore Mine in 1970. John Norton operates a Joy Loader (mechanical shovel), which was manufactured by Joy Manufacturing Co. (UK) Ltd, who were also the pioneers of the concept of continuous mining and led the way in designing machines to speed the cutting, loading and transportation cycle of underground mining. Llanharry Iron Ore Mine closed in the 1970s.

Strikers' Pit in 1926. Two striking miners at the top of a very small pit they had opened to obtain coal for their own domestic use during the strike. The small pit is situated behind the boy in the photograph. The winch, which was hand operated, was assembled and made by the striking miners. The narrow, small shaft of the pit would probably be approximately 10ft in depth. The pit was situated at Brynna Woods near Llanharan in 1926.

Strikers' Pit in 1926. The strikers have sunk a small pit on an outcrop at Brynna Woods near Llanharan. The wheel of the upturned bicycle was the winding pulley and the narrow, small shaft of the pit would also probably be approximately 10ft in depth. The coal seam worked by these small pits was the Cribbwr Fach. There were many of these small strikers' pits sunk at Brynna Woods during the 1926 strike.

Remains of Llanbad Colliery, near Brynna, in 1950. Some underground smells were unique to coalmining. The thick clouds of coal dust in the return airway; the sweat from the toil of the miner; the pit ponies at work hauling the drams; peeling a fresh orange; the fumes following shotfiring; a miner relieving his bowels and throwing it in the gob or behind the lagging of a ring with his size four shovel. There were no toilets underground and the gob is an area where the coal has been mined. Llanbad Colliery was abandoned between 15 July and 20 December 1916.

South Rhondda Colliery, near Llanharan, East of Ogmore Valley, in 1900. South Rhondda Colliery worked the Rhondda No.2 and Rhondda No.3 coal seams. The colliery was opened in 1898 by the South Rhondda Colliery Company Ltd and was closed when acquired on Vesting Day in 1947 by the NCB.

Mill Colliery, Cefn Cribbwr, near Kenfig Hill, in 1910. Mill Colliery was owned by the Bryndu Colliery Co. In 1910 the colliery employed 253 miners. Seams worked were: the Two-Feet-Nine; Six-Feet; Four-Feet (Esgyrn); Rock Seam; Crow Foot; and Red Vein (North Fawr). Mill Colliery was abandoned on 29 February 1912.

Bryndu Drift Mine, Cefn Cribbwr, near Kenfig Hill, in the 1930s. Section of coal from Garw Seam to neighbourhood of Six-Feet in North and South Drifts. Its sequence was broadly similar to that of Newlands Colliery. Bryndu Drift Mine was worked in the nineteenth century by the Bryndu Coal and Coke Co. to feed the Cefn Iron Works. At approximately 5.00 p.m. on Sunday 27 May 1858, eleven miners were killed by an explosion. Bryndu Drift Mine was abandoned on 31 January 1912.

Tytalwyn Drift Mine, Kenfig Hill, in 1915. Tytalwyn Drift Mine was owned by Ton Phillip Rhondda Colliery Co. Ltd and became part of Aberbaiden Colliery in 1906. In 1913 the colliery employed 261 miners. I, like many others, miss the old familiar sound of the colliery hooter (hwtar). The hooter is a siren blowing from the colliery in the village, informing the relevant shift coming on of the time. Tytalwyn Drift Mine was closed in January 1959 by the NCB.

Aberbaiden Slant Mine in 1919. Aberbaiden Slant Mine, Kenfig Hill, was sunk in 1906 and finally opened in 1908 by Messrs Baldwin Ltd on the site of an older slant (the mine roadway turned at an angle from the main roadway, to the dip of a seam) and consisted of the Aberbaiden Slant, to a depth of 1,500 yards, and the Pentre Slant, sunk to a depth of 1,900 yards. The mine was opened to work the Rock Fawr seam, at a section of: coal 24in; dirt 12in; coal 24in. The coal seams worked were the Rock Fawr and the Hafod. Aberbaiden Slant Mine was closed in January 1959 by the NCB.

Left: Pentre Slant Mine, Kenfig Hill, in 1910. The colliery was opened in 1908. The Slant along with Aberbaiden and Newlands worked this difficult area of the south crop. The owner of the Slant prior to nationalisation was Baldwyn's Ltd. In 1954, with a manpower of 377, it produced 67,000 tons. The coal seam worked was the Rock Fawr. *Right*: Pentre Slant Mine in 1920. Pentre Slant Mine was closed on 6 November 1959 by the NCB.

Using water on a puncher (pneumatic pick) in 1958. In the photograph, the heading man can be seen ripping down the coal in a coal heading (roadway) using a puncher (one of the many causes of vibration white finger) in preparation to erect steel arches (rings), which are then lagged with timber erected above and around the sides of the steel rings. These would ensure support for the roof and sides and also no stones could fall on a miner passing by. The gate road was advanced on the afternoon and night shifts. The gate road in a longwall face, usually on the air intake side, is a heading for the transporting of coal from the coalface, normally by conveyor belt. It also provides width for a haulage transport system, for the supplying of steel arches, timber, etc., for the advancing of the conveyor head. Most electrical equipment was in the gate road in a district.

Newlands Colliery, near Pyle, in 1952. The measures from the Gellideg (Cribbwr Fawr) seam to the Caerau Caegarw are described on the following section, which is a composite one made up of the Main Drift, No.1 Scheme, Upper Nine-Feet, South Fawr to Caerau and the No.3 Drift, Gellideg to Upper Nine-Feet. The depth of shaft was 759ft 7in. Newlands Colliery Upcast Drift was opened in 1951 by the NCB. In 1967 the Newlands manager was D.W. Jones (2,918 First Class) and the undermanager was L. Bradley (7,731 First Class).

Newlands Colliery in 1952. The colliery was opened in 1918 by Newlands Colliery Co. Ltd and the Cribbwr Fawr Collieries Ltd. According to the NCB list of mines taken over on Vesting Day 1 January 1947 the colliery was opened in 1918, but this information is doubtful as there was a mine on the site long before this date. There was also a considerable water problem in some districts of the mine. Newlands Colliery was closed on 22 March 1968 by the NCB.

Newlands NUM Lodge Banner in 1957. The following are the findings of the exploratory boreholes for the future Margam Super-Pit. Margam Park No.1 Borehole was 286ft OD and sited 2,050yds E 14° S of St Mary's Church, Margam Park. Its average dip was 15°. It was drilled in 1954. Seams were: Cefn Coed Marine Band at 207ft 9in; Britannic Marine Band at 296ft 10in; Hafod Heulog Marine Band at 332ft, repeated at 370ft 10in; Two-Feet-Nine at 551ft 7in; Upper Four-Feet at 595ft 7in; Lower Four-Feet at 630ft 7in; Upper Six-Feet at 723ft 2in; Red Vein at 789ft 1in; Upper Nine-Feet at 835ft, Lower Nine-Feet at 873ft 3in; Bute at 1,013ft 3in; Amman Marine Band at 1,053ft 1in; Yard at 1,175ft 6in; Lag Fault at about 1,225ft, Gellideg at 1,342ft 6in; Garw at 1,503ft 6in. The base of coal measures was 2,650ft and it was drilled to 3,550ft.

Margam Park No.2 Borehole was 320ft OD and it was sited 1,070yds E 30° N of St Mary's Church, Margam Park. It had an average dip of 15°. It was drilled in 1954-1955. Coal seams were: Lower Cwmgorse Marine Band at 105ft 6in; Five Roads Marine Band at 223ft; Foraminifer Marine Band at 295ft 7in; Pentre at 346ft 3in; Britannic Marine Band at 494ft; Hafod Heulog Marine Band at 528ft 6in; Two-Feet-Nine at 736ft 6in; Upper Four-Feet at 833ft 8in; Lower Four-Feet at 874ft 10in; Six-Feet at 994ft 10in; Caerau at 1,018ft 10in; Red Vein at 1,084ft 6in; Upper Nine-Feet at 1,152ft 9in; Lower Nine-Feet at 1,221ft 7in; Lag Fault at about 1,370ft; Lower Five-Feet at 1,430ft 7in; Gellideg at 1,580ft; Garw at 1,747ft 7in, drilled to 2,029ft 8in.

Margam Park No.3 Borehole was 359ft OD. It had an average dip of 10°. It was drilled by percussion to 389ft 3in; the record to that depth being of an earlier borehole on the same site. It was drilled in 1954-1955.

Margam Park No.4 Borehole was 327ft OD. It had an average dip of 15°-20°; it was drilled in 1955. Coal seams were: Hafod Bottom Coal at 163ft 4in; Lower Cwmgorse Marine Band at 286ft 11in; Abergorki Top Coal at 366ft 10in; Abergorki Bottom Coal at 401ft 5in; Foraminifer Marine Band at 488ft 1in; Cefn Coed Marine Band at 855ft 7in; Hafod Heulog Marine Band at 1,050ft; Two-Feet-Nine at 1,254ft 10in; Upper Four-Feet at 1,306ft 6in; Lower Four-Feet at 1,348ft 9in; Six-Feet at 1,453ft 4in; Caerau at 1,534ft 9in; Red Vein at 1,555ft 5in; Upper Nine-Feet at 1,602ft 1in; Lower Nine-Feet at 1,638ft 9in; Bute at 1,670ft; Amman Marine Band at 1,702ft 4in; Yard at 1,768ft 1in; Middle Seven-Feet at 1,788ft 4in; Lower Seven-Feet at 1,856ft 6in; Upper Five-Feet at 1,974ft 8in; Lower Five-Feet at 2,136ft 6in; Gellideg at 2,286ft 3in; Garw at 2,495ft 4in. It was drilled to 2,530ft 2in.

In everlasting memory to the miners who lost their lives in Ogmore Vale and the Llynfi, Garw and Ogwr valleys

Date	Mine	Fatalities
11 November 1852	Bryndu	Collier Thomas Jenkin (35) killed by an explosion.
24 December 1863	Duffryn	Fifteen miners killed by afterdamp.
25 March 1868	Oakwood	Haulier John Morris (18) killed by drams.
21 February 1874	Bryndu	Miner John Harries (36) killed by an explosion.
19 July 1884	Coegnant	Collier William Rees (16) killed by a fall of coal.
3 July 1887	Garw	Collier Thomas Jones (14) killed.
11 August 1887	Wyndham	Repairer John Snow (25) killed.
14 January 1888	Meiros	Collier David Morgan (26) killed.
20 May 1888	Garw	Haulier Phillip Morris (25) killed.
18 May 1889	International	Collier David Jones (18) killed.
11 July 1889	Ffaldau	Miner George Robbins (20) killed.
25 March 1890	International	Collier David Reynolds (49) killed by a fall of clift from the front of the rippings in the Two-Feet seam, which he had been told to take down.
16 October 1890	Darren	Collier Michael Howells (21) injured by fall of roof in the Daren Rhestyn seam and died on 11 March 1891. The pillar and stall method was used.
29 October 1890	Western	Collier Henry Roach (38) killed by a fall of side from the edge of the rippings in the Six-Feet seam.
10 December 1890	Western	Stoker James John (23) died on 15 December. He was taking an empty truck down an incline from the boilers, when a brake-van was drawn up against the truck and he was knocked down and injured.
29 January 1891	Meiros	Pumper John Barker (13) killed by a fall of rock.
5 August 1891	Caerau	Stoker Stephen Overston (50) died on 13 August. He was going up a ladder to the top of the boilers; he spilled some oil from a 'Comet' naked flame lamp on his clothes and while oiling the mechanical stoker his clothes caught fire at the lamp.
12 June 1893	Garw	Collier John Jones (21) and Collier David Jones (18) were killed by a loaded dram, which had been left standing on an inclined road and running back wild into their stall, but how it came to do so was not satisfactorily ascertained. There appeared to have been carelessness in the changing of the drams.
15 June 1893	Western	Collier Alfred Thomas (28) was killed by a fall of roof on a road in the Six-Feet seam, while resting on the side of the heading. The place was said to have been well timbered.
28 September 1894	South Rhondda	Collier's boy T.B. Butt (17) was killed by a fall of roof (rock) at the face of the stall in the No.3 seam. The fall was fully 100ft in area, from 2ft to nothing in thickness. Five props were supposed to have been stood under it. The pillar and stall method was in use.
18 August 1899	Llest	Nineteen miners killed by an explosion.

Avon (Afan) Colliery, Blaengwynfi, Afan Valley, Glamorganshire, in 1950. The South Pit was sunk to 1,504ft 4in in 1877 by Sir Daniel Gooch. On 7 November 1892 the accident reports show that twenty-one-year-old shoeing smith David Simeon was killed when he was kicked by a horse in the stables.

Avon Colliery in 1955. North Pit was 954ft. Seams worked were: Hafod (Bottom Coal) seam at 538ft 6in; Abergorki (Bottom Coal) at 707ft 6in; Pentre Rider at 786ft; Pentre at 853ft 6in; Gorllwyn at 1,023ft 6in, sunk to the western branch of the Glyncorrwg Fault to Lower Nine-Feet at 1,269ft 11in; Bute at 1,295ft 2in; Amman Rider at 1,342ft 8in; Yard at 1,413ft 10in; Middle Seven-Feet at 1,521ft 5in, sunk to 1,528ft 2in.

Left: Avon Colliery mechanical engineers and craftsmen (fitters) working on the pit sheaves in 1955. In 1905 it was owned by the Great Western Railway Co. and from 1909 until nationalisation it was owned by the Ocean Coal Co. In 1909 the colliery employed 907 miners; in 1958 with a manpower of 634 it produced 141,691 tons and in 1961 with a manpower of 533 it produced 116,722 tons. In 1967 the Avon manager was J. Smith (6,390 First Class) and the undermanager was A. Davies (6,068 First Class). Avon Colliery was closed in September 1969 by the NCB.

Ynyscorrwg-Blaengwynfi Colliery in 1920. Seams worked were the Rhondda No.1 Rider at 312ft 6in and Rhondda No.2 at 680ft 4in, sunk to 712ft 3in. In the 1920s the colliery was owned by the Glyncorrwg Colliery Co., who worked the Rhondda No.2 seam extensively. Ynyscorrwg-Blaengwynfi Colliery was opened in 1881 and was closed when acquired on Vesting Day 1947 by the NCB.

Corrwg Rhondda Colliery, near Glyncorrwg, in the 1930s. Corrwg Rhondda Colliery, locally known as the Tunnel Colliery, was opened in 1881. On 16 May 1898 the accident reports show that forty-two-year-old collier John Chappel was fatally injured. In 1913 the colliery employed 172 miners. In the 1920s it was owned by the Glenavon Garw Co. Corrwg Rhondda Colliery was closed when acquired on Vesting Day 1947 by the NCB.

Glyncorrwg Colliery in 1953. Seams worked were: Daren-Rhestyn horizon at 40ft; Rhondda No.2 horizon at 221ft 5in, sunk through eastern branch of Glyncorrwg fault to Abergorki at 467ft; Pentre Rider at 546ft 2in, sunk through reversed fault of 133ft to Pentre Rider at 679ft; Lower Pentre at 721ft 9in; Gorllwyn at 821ft 2in; Two-Feet-Nine at 1,080ft 5in; Four-Feet at 1,125ft 1in; Six-Feet at 1,263ft 1in. The total depth was 1,283ft 1in. Glyncorrwg Colliery was sunk in 1867 by Glyncorrwg Collieries Ltd. In 1910 the colliery employed 567 miners.

Glyncorrwg Colliery underground haulage driver Jack Edwards in the 1930s. Glyncorrwg Colliery, South Pit, was sunk to 1,222ft. From 1912 to 1919 the North Pit remained idle when the South Pit employed 144 miners and worked the Red Vein seam. In 1926 the colliery was owned by Glyn Neath Collieries Ltd. In 1927 the manpower was 384 working the Nine-Feet and the Peacock seams. In 1928 it was owned by the Amalgamated Anthracite Collieries Ltd.

First day at the pit for Glyndwr 'Glyn Bobby' Thomas in 1928, aged fourteen. Glyn did many types of underground work; he served as a shotsman, deputy overman, and senior overman and also acted as undermanager and manager of Glyncorrwg Colliery. He was involved in an explosion at the colliery on 13 January 1954 where he was in charge of the unit and was highly commended for the organisation of the withdrawal of all workmen and the rescue of the injured. In July 1972 he began the formation of the South Wales Miners Museum at Afan Argoed, to portray the history and mining communities of the Afan Valley. Glyndwr 'Glyn Bobby' Thomas (1914-2001) will always be remembered for dedicating his time to the community of upper Afan Valley. Glyncorrwg was the last working colliery in the Afan Valley. Glyncorrwg Colliery was closed on 1 May 1970 by the NCB.

North Rhondda No.1 and No.2 Colliery lamp check. North Rhondda Colliery was opened in 1908 by the North Rhondda Colliery Co. to the Rhondda No.2 seam. On 12 December 1913 the accident reports show that sixty-four-year-old repairer John Jones was fatally injured. In 1947 the colliery employed 160 miners; in 1956 with a manpower of 257 it produced 75,832 tons, and in 1957 with a manpower of 238 it produced 64,866 tons.

North Rhondda No.1 and No.2 Colliery in 1959. The two miners at work underground in 1959 are both using a Sylvester. This popular and handy tool was used in a variety of collieries in the South Wales Coalfield and was also called a Puller, Buller, Crongy, Dragger and Faker. A Sylvester was a ratchet type of pulling appliance for the dragging of heavy equipment in a colliery. The tool was totally banned from use by 1987, being regarded as unsafe, after about 50 years of common usage in most coalmines throughout the coalfield. It had a long history of accidents including fatalities. North Rhondda No.2 Colliery was closed in May 1947 and No.1 Colliery was closed in July 1960 by the NCB.

Nantewlaeth Colliery Rescue Team in 1920. Nantewlaeth Colliery was owned by Gibbs Navigation Collieries Ltd prior to nationalisation. In 1947 the colliery employed 536 miners and worked the Rhondda No.2 and Six-Feet seams. Nantewlaeth Colliery was closed in November 1948 by the NCB.

Glyncymmer Colliery, Cymmer, in 1920. Glyncymmer Colliery was sunk in 1895 by Gibbs Collieries and Glenavon Garw Collieries Ltd. On 11 February 1903 the accident reports show that fifty-three-year-old labourer John Thomas was fatally injured and that on 16 August 1906 seventeen-year-old labourer David Thomas Hopkins was also fatally injured. In 1913 the colliery employed 170 miners. Glyncymmer Colliery was closed when acquired on Vesting Day 1947 by the NCB.

Left: Building Cynon Colliery in 1908. Cynon Colliery was opened by Cynon Colliery Ltd in 1909. In 1910 the colliery employed 800 miners. *Right*: On top of the pit at Cynon Colliery in the 1930s. In 1913 the colliery employed 500 miners and in 1947 the manpower was thirty-one. On 15 December 1914 the accident reports show that thirty-seven-year-old collier Benjamin Evans was fatally injured. Cynon Colliery was closed when acquired on Vesting Day 1947 by the NCB.

Duffryn Rhondda Colliery, Cymmer, in 1930. Duffryn Rhondda Colliery, No.1 Shaft was 508ft OD. Coal seams worked were: Wernddu at 489ft; Tormynydd at 919ft; Clay at 1,100ft; Albert at 1,282ft 6in; Victoria at 1,360ft 6in; Two-And-A-Half at 1,526ft 1in; Caedavid at 1,556ft 9in. Duffryn Rhondda Colliery, No.2 Shaft 45yds east of No.1 Shaft, deepening from Four-Feet. Duffryn Rhondda Colliery was closed in November 1966 by the NCB.

An underground telephone that was in use at several collieries in the 1970s. This rather primitive communication apparatus was operated by lifting the earpiece from its holder on the left in the photograph and placing it to the ear, then turning the handle (right) with great speed would ring the bell (top) on the telephone at the other end and by shouting in the mouthpiece (centre). The person at the other end on his telephone would, hopefully, be able to hear what the person was saying. This telephone in its final years was used on pit bottom.

The Countryside Centre and Miners' Museum in 2002. An exciting hands-on exhibition graphically demonstrating the landscape and history of the Afan Valley is the centre-piece of the Countryside Centre. The social history of the valleys' mining communities is portrayed in the South Wales Miners' Museum.

Canolfan Cefn Gwlad ac Amgueddfa'r Glowyr, 2002. Yr arddangosfa cyffrous, ymarferol sy'n dangos y tirwedd a hanes Cwm Afan, yw canolbwynt Canolfan Cefn Gwlad. Darlunir hanes gymdeithasol cymunedau glo y Cymmoedd yn Amgueddfa Glowyr De Cymru.

Morfa Colliery, Taibach, Port Talbot, in 1863. Morfa Colliery was sunk in 1847 to supply coal to the Taibach Copper Works. The colliery was owned by C.R.M. Talbot Esq. and leased to Messrs Vivian & Sons. Morfa had an unfortunate history of explosions and it became known as the pit of Ghosts. At approximately 1.00 a.m. on Thursday 26 November 1858, an explosion killed four miners. At approximately 10.00 a.m. on Saturday 17 October 1863, another explosion killed thirty-nine miners. On this fateful day, night overman John Evans and William Dommer reported the district to be free of gas and 400 men went down to work the dayshift. William Grey the manager followed them at about 9.30 a.m. and undermanager William Barrass at 10.00 a.m. The cage carrying the undermanager gave an ominous shudder and he guessed the worst. The blast shattered the west side of the old Nine-Feet district, where there were said to be numerous abandoned workplaces in which gas, quite commonly, accumulated. Despite their reputation, naked flame lights were still in use as late as 1870.

On Monday 14 February 1870 an explosion killed twenty-nine miners. The shafts at the colliery were being deepened to the lower seams. A keg holding about 40lb of blasting powder was kept in an engine house near pit bottom. It was ignited here, so management asserted that it was responsible for the deaths on this day of twenty-nine or thirty, but only twenty-five were named. Others said that the effects of such a blast could not have been felt at the stables about 600yds distance inbye ('he has gone inbye', means he has gone towards the coalface). All of the thirty-two horses kept there were found dead. About ten days after the blast, men at work clearing the falls and erecting new timber supports were using naked flame lights; one of them, a lad named James Badge, was observed walking to join a gang, with a naked torch in his hands. There came an ignition of gas at his torch 'and the hand holding the flaming flambeau was almost entirely burnt away.' The manager at this time was Mr Gay. At the inquest he said that on the 6 February a patch of coal in the Nine-Feet seam had taken fire and the shift overman on duty thought it was not worth reporting. He was dismissed for neglecting his duties. On 10 March 1890 another explosion killed eighty-seven miners. At this time Morfa Colliery was producing 400 tons of coal a day from the Nine-Feet and the Cribber seams. On one Tuesday morning a little after midnight there came a violent explosion in the latter of these districts that was to take eighty-seven men and boys' lives. Safety flame oil lamps were in use. Morfa Colliery was abandoned in 1914.

'The Flint Lamp' was found during opencast mining at Hirwaun, near Glyn Neath, during the 1970s. The lighting device known as Speddings Flint and Steel Mill was invented c.1747-1748 by Mr Carlyle Spedding, a manager at Whitehaven Pits, Cumberland. A piece of flint was held against the steel disc which turned at a rotation of about four to one. A shower of sparks was produced and was said to produce enough light for two or three miners to work by. Usually the apparatus was worked by children as young as four years of age. It was considered safe in a gaseous atmosphere. This was soon proved quite wrong when several serious colliery explosions were caused by its use.

Craig-Y-Llyn Level, Rhigos, near Glyn Neath, Vale of Neath, Glamorganshire, in 1996. This photograph shows the main entrance and intake (the route taken by fresh air from the surface to the workings). The level worked the Rhondda No.2 bituminous coal seam. During its coaling life there were many abandonments. Craig-Y-Llyn Level has a current operating licence which was granted on 4 April 1995.

Remains of Rock Colliery No.2 Drift Mine, Glyn Neath, in 1975. Rock Colliery was opened in 1943 by the Rock Colliery Co. and was owned by the Amalgamated Anthracite Collieries Co. prior to nationalisation. In 1954 the colliery employed 636 miners and produced 186,000 tons and in 1960 with a manpower of 325 it produced 85,000 tons. Rock Colliery No.2 Drift Mine was closed on 23 October 1961 by the NCB.

Left: Remains of Pontycoed Drift Mine, Glyn Neath, in 1962. Pontycoed Drift Mine was opened in 1888 and abandoned in 1907. *Right*: Lyn Drift Mine, Glyn Neath in the 1980s. Lyn Drift Mine was owned by the Rhondda Mountain Coal Co. in 1878. In 1967 it was owned by Lyn Colliery Co. and worked the Rhondda No.1 and Rhondda No.2 seams. Lyn Drift Mine was abandoned on 21 July 1995.

Pentreclwydau No.1 Drift Mine, near Glyn Neath, in 1967. In 1960 Pentreclwydau No.1 Drift Mine was owned by the NCB, employed 352 miners and produced 85,490 tons of coal. In 1961 the colliery employed 513 miners and produced 85,490 tons. In 1967 the Pentreclwydau manager was J.H. Ellis (4,931 First Class) and the undermanager was H. Lloyd (8,648 First Class). Pentreclwydau No.1 Drift Mine was closed on 27 May 1967 by the NCB. Pentreclwydau South Mine was abandoned on 28 July 1988.

Blaengwrach Drift Mine, near Glyn Neath, in 1950. Blaengwrach Mine dates back to workings in 1814 (No.1 Level 1814, No.2 Level 1814 and No.3 Level 1924). In 1962 Blaengwrach New Mine was formed by the NCB near Aberpergwm Washery. In 1967 the Blaengwrach manager was T.A.G. Martin (3,939 First Class) and the undermanager was T.R. Evans (7,923 First Class). In 1981 the districts worked were the N101 and the N102 in the Nine-Feet seam.

Blaengwrach Drift Mine Locomotive in 1982. The N101 was 153yds in length with a section of 5ft 5in; daily advance of 8ft; daily shifts of two; coal cutter ranging drum shearer; roof supports were self advancing; daily output of 590 tons; expected life distance was 563 yards.

Left: Blaengwrach Colliery in 1984. N102 was 177yds in length with a section of 5ft 3in; daily advance of 8ft; daily shifts were two; cutter ranging drum shearer; roof supports were self advancing; daily output of 639 tons; manshifts per day on development were forty; manshift per day on coalface was sixty; elsewhere below ground it was eighty; manshifts per day on surface was forty, making the total 220. Blaengwrach Drift Mine was closed in 1983 by the NCB. *Right:* Empire Colliery, Cwmgwrach, near Glyn Neath, in 1947. The colliery was opened in 1903. Empire Colliery was closed when acquired on Vesting Day 1947 by the NCB.

Underground double drum main and tail electric haulage in 1910. The haulage engine, a steam, compressed air or electrical type of fixed engine, was used on the surface and below ground. It was used for taking into a district supplies for the face and returning to the pit bottom with a full journey of coal. The main rope is attached to a journey of drams to pull the drams outbye (towards the shaft or to the mouth of a level). The tail rope is attached to a journey of drams, to pull the drams inbye; when the gradient is unsuitable for the free running of it towards the coalface, a terminal sheave would control the tail rope.

Underground electric generator station in 1910. The generator supplied electricity to underground machinery including pumps and haulage engines. The haulier, pronounced 'halier', is a miner who drives a horse to the coalface or stall with an empty dram and returns to the 'double-parting' with a full dram of coal. In a conveyor face system he would transport supplies to the supply road. He was in sole charge of the horse.

Aberpergwm Drift Mine, near Glyn Neath, on 31 May 1972. Aberpergwm Drift Mine was first sunk in 1863. Conveniently located between the coast and the early iron deposits, the coal outcrops of Glyn Neath had been the scene of mining activity for close on 300 years. Aberpergwm mine dates back to 1863, though part of the workings coincides with those of even older collieries. In its early years, the colliery was known as Pwllfaron Slant. Over the years, coal has been taken from the Eighteen-Feet (Six-Feet), Four-Feet, Nine-Feet, Three-Feet (Peacock) and Cornish seams.

In 1967 the Aberpergwm manager was N.D. Lewis (5,582 First Class) and the undermanager was J.A. Griffiths (7,924 First Class). In 1969, several geological problems threatened the pit's future and an exploratory heading was driven into the Eighteen-Feet seam, to assess pillars of coal left by earlier pillar and stall methods used 100 years before. These pillars held sufficient high-quality anthracite to justify the driving of a new 899ft drift and immediate investment in new machinery and transportation systems. Six years later, in 1975, this rich area of coal began to show signs of running out. However, the mine's performance persuaded the NCB to invest another £725,000 in a scheme to reach further reserves lying to the northeast. Twin cross-measures were driven right through the notorious Pentreclwydau Fault, revealing an estimated fifteen to twenty years of prime anthracite, in the Eighteen-Feet, Nine-Feet and Three-Feet seams.

In 1976 the new drivages reached the Nine-Feet and test headings were pushed out to see if geological conditions would allow a completely new mechanised longwall face to boost output from the mine. The NCB wrote at this time: 'If all continues to go well, Aberpergwm could still be producing valuable coal supplies in the year 2000, almost 140 years after production first began.' The weekly output averaged around 3,000 tons and was conveyed directly across the A465 Heads of the Valleys Road, to a central coal preparation plant (washery) shared with neighbouring Blaengwrach Colliery. Washed coal was then shipped out by road to their eventual destinations. The Aberpergwm Drift Mine, with a manpower of 351 in the pit and 102 in the washery, produced an annual saleable output of 100,000 tons. It produced an average weekly saleable output of 3,000 tons; average output per man/shift at the coalface 5 tons; average output per man/shift overall 1 ton 7cwt; deepest working level 1,800ft; number of production headings working were four.

Aberpergwm Drift Mine on 31 May 1972. Aberpergwm Drift: length 2.2 miles; gradient 1 in 7 to 1 in 4; manriding capacity per spake (train journey) forty; coal conveying capacity per hour from pit 400 tons; average weekly washery throughput 6,000 tons; type of coal was anthracite; markets were domestic, industrial power stations and export; total capital value of plant and machinery in use £1.2 million and the estimated workable coal reserves were 6.7 million tons. The licensee in 2002 was Anthracite Mining Ltd, Rheola Works, Resolven. Aberpergwm Drift Mine is coaling (producing coal) today, May 2002.

Glyn Castle Colliery, Resolven, in 1960. Glyn Castle Colliery was sunk in 1875. The colliery was owned by Cory Brothers Co. Ltd prior to nationalisation. In 1955 the colliery employed 266 miners and produced 65,836 tons of coal. In 1958 with a manpower of 266 it produced 67,292 tons and in 1961 with a manpower of 279 it produced 33,703 tons. Glyn Castle Colliery was closed in 1965 by the NCB.

Main No.1 Colliery, near Neath, in 1900. Main Colliery was opened in 1860 by the Neath Abbey Coal Co. It was also known as Bryncoch & Pwll Mawr. Main No.1, No.2, No.3, No.4, No.5 and No.6 Colliery past owners include the Neath Abbey Coal Co., Dynevor United Collieries Co. Ltd and the Main Colliery Co. Ltd. In 1913 the No.1 Colliery employed 834 miners.

Main No.3 and No.4 Colliery in 1900. In 1913 the No.3 and No.4 employed 560 workers. In the photograph the winding engine houses can be seen at the rear of the colliery headframes. The winding engine provides motive power for the drum around which is wound a rope attached to the cages travelling the pit shaft and is the key for conveying men, minerals and every single item of material and machinery into and out of the pit.

Main No.5 and No.6 Colliery in 1900. On 15 November 1860 the accident reports show that an eleven-year-old door boy was crushed between two drams and on 26 October 1868 they show that thirty-three-year-old collier J. Lewis was killed by suffocation. Main Colliery was abandoned in 1929.

Court Herbert Colliery, near Neath, in 1914. Court Herbert Colliery was opened in 1854 by Richard Parsons. On 24 September 1868 the accident reports show that thirty-five-year-old collier Thomas Jones was killed by a fall of roof and on 20 July 1892 that thirty-six-year-old collier Jas Shaddick was killed by a fall of clod from the side, for the want of sprags (supports). In 1913 the colliery employed 352 miners and in 1918 the manpower was 371. The colliery was also owned by Dynevor Coal Co. and the Main Colliery Co.

Court Herbert Colliery underground furnaceman in 1897. Furnace ventilation is an obsolete method of ventilation in which a fire is kept burning near the bottom of the upcast shaft, to draw air into the mine workings. A dumb drift was the alternative method of furnace ventilation, where a drift was made from a place a suitable distance from the bottom of the upcast shaft. This drift is inclined and joins the upcast shaft. The furnace is placed at the lower end of the dumb drift and is fed with fresh air from the downcast shaft or by any other suitable method. The hot air rising from the dumb drift passes into the upcast shaft and draws to the surface the foul air from the vicinity of the pit bottom.

Court Herbert Colliery screens in 1913. The screens were situated on the colliery surface where coal was sorted into different sizes and any rubbish including stone and timber in the coal was picked out. On 7 March 1894 the accident reports show that sixteen-year-old haulier Thomas Thomas was killed when he was crushed between empty drams while attempting to detach the horse's chains before the dram ceased running. Court Herbert Colliery was abandoned in 1929.

Crugau No.2 Level, near Neath, in 1965. Crugau Level was owned by the Crugau Colliery Co. Ltd and Rhys Jeffreys. The photograph shows the haulier giving his horse a helping push, always working together as a team. Crugau No.2 Level was abandoned on 6 March 1993.

Oakland Colliery, near Neath, in 1920. Oakland Colliery was opened in 1914 by the Oaklands Colliery Co. Ltd. The wooden headframe of this short life pit had been situated some distance from the winding engine house, indicating a shallow pit to mine the upper coal seams. Oakland Colliery was closed in 1953 by the NCB.

Students cogging and gobbing competition, Vale of Neath, in 1906. A cog (cogyn) is a roof support that was built in the gob (the waste area left after the removal of coal), which helped to control roof conditions. The cog was constructed of interlaced horizontal wooden pieces with the alternate layers at right angles laid from floor to roof and filled with rubble. Packs (stone-built walls) were also erected to control the roof and ventilation conditions.

Students timbering competition, Vale of Neath, in 1906. The students are standing timber known as the collar and arms using the Welsh Notch at the joints. One of the better and favourite hatchets, which were used, was Y Bwyall Tŷ Gwyn (the white house hatchet). The miners always kept their Tŷ Gwyn hatchets sharp as razor blades. One hundred students took part in the 1905 competition and the highest colliery officials acted as judges.

Darren Level, Neath, in 1967. The collier seen in the photograph is a skilled miner who is cutting the coal from the coalface. Darren Level was abandoned on 31 March 1972.

In 1584 the first copper smelting works were at Neath, using imported Cornish ore and locally mined coal. In 1898 a new dock was opened at Port Talbot to serve the coalfields.

Neath No.9 area general manager in 1961 was C. Round; the area production manager was P.G. Weeks. The group managers for the area were: No.1, H. Jarman; No.2, T. Walters; No.3, David Evans; No.4, E.G. Maggs. The No.1 group of collieries were: Blaenant, Cefn Coed, Glyncastle, Ffaldyre. The total output was 267,704 tons; total manpower 1,285. The No.2 group of collieries were: Onllwyn No.1, Onllwyn No.3, Seven Sisters, Dillwyn. Their total output was 332,371 tons; total manpower 2,183. The No.3 group of collieries were: Abercrave, Pwllbach, Varteg, Yniscedwyn, Tirbach. They had a total output of 310,819 tons and a total manpower 21,746. The No.4 group of collieries were: Cwmgorse, East, Ammanford, Pantyffynnon, Wernos, Abernant. Their total output was 485,647 tons; total manpower was 2,838.

In everlasting memory to the miners who lost their lives in the Vale of Neath

Date	Mine	Lives Lost
1758	Wernfraith	Ten miners killed by an explosion including manager and owner William Shepley.
1764	Winch Pond	Eighteen miners killed by an explosion.
12 May 1795	Fire Engine	Twenty miners killed by an explosion.
9 June 1820	Cwmgwrach	Five miners killed by an explosion – including two girls, Elizabeth Pendry (6) and Annie Tonks (12).
May 1837	Eaglebush	Eight miners killed by an explosion.
29 March 1848	Eaglebush	Twenty miners killed by an explosion.
6 January 1853	Garasddu	Collier Llewellyn Rees (69) killed by a fall of stone.
5 April 1859	Chain	Twenty-six miners killed by an explosion.
12 April 1860	Maesmarchog	Haulier killed by falling under drams at level.
20 October 1860	Bryncoch	Sinker Evan Jones (24) fell into sinking shaft.
3 March 1868	Bryndowy	Haulier Jonathan Edwards (14) killed by drams.
10 July 1868	Bryndowy	Haulier Thomas Jones (27) killed by the kick of a horse.
16 September 1868	Pwllfaron	Collier T. Richards (40) killed by a fall of coal.
23 February 1872	Gnoll	Sinker H. Mathews killed by a broken rope.
2 July 1872	Dynevor	Collier J. Bowen (15) killed by a fall of stone.
11 July 1872	Brithdir	Haulier L. Thomas (17) killed by drams.
16 October 1872	Resolven	Collier D. Price (35) killed by a fall of stone.

Onllwyn No.1 Drift Mine, Onllwyn, Dulais Valley, Glamorganshire, in 1900. Onllwyn No.1 Drift Mine was opened in 1845 by Evans & Bevan. On 11 January 1876 the accident reports show that a blacksmith was killed by an explosion of steam blowing a piece of iron violently against the deceased, whilst he was engaged in repairing a small piston. In these anthracite valleys the coal industry developed at a much slower pace.

Onllwyn No.1 Drift Mine timber yard in 1900. On 27 November 1898 the accident reports show that sixty-three-year-old labourer Howell Powell was killed when he was knocked down by a loaded dram on a short surface incline. In 1954 the colliery employed 628 miners; in 1955 with a manpower of 637 it produced 114,475 tons and in 1956 with a manpower of 629 it produced 104,956 tons. Onllwyn No.1 Drift Mine was closed in April 1964 by the NCB.

Seven Sisters Colliery, Seven Sisters (Blaendulais), in 1920. Seven Sisters Colliery was opened in 1870 by David Bevan. In 1872 the small hamlet of Blaendulais was renamed Seven Sisters after the seven daughters of David Bevan, the local coalowner. On 29 July 1875 the accident reports show that thirty-year-old sinker E. Morris was killed by falling down the shaft.

Seven Sisters Colliery in 1942. On 3 October 1886 the accidents reports show that thirty-eight-year-old repairer David Davies was killed by a fall of stone while engaged in repairing an airway and on 25 June 1888 forty-five-year-old collier D.H. James was killed by an explosion of firedamp. In 1956 with a manpower of 685 it produced 124,326 tons. Seven Sisters Colliery was closed on 4 May 1963 by the NCB.

Cefn Coed Colliery Upcast Shaft, Crynant, in 2002. The Neath Colliery Co. commenced sinking a pair of shafts in 1921 on the east bank of the Dulais River, but the sinking was suspended until 1924. In 1928 the Amalgamated Anthracite Collieries acquired Cefn Coed Colliery and completed the work. The Big Vein was found at a depth of 2,250ft and the Peacock Seam was reached in 1931. The main seams the colliery worked were the Dulais, Peacock, Nine-Feet and White Four Feet.

Cefn Coed Colliery winder in 1968. Cefn Coed Colliery closed in 1968 when a new drift mine to be known as Blaenant Mine was opened, which still retained the use of the two steam winders on the old Cefn Coed shafts. A Worsley Mesnes steam-winding engine was situated on the downcast shaft with a Markham steam-winding engine on the upcast shaft. The Worsley Mesnes steam-winding engine still stands on the downcast shafts in its engine house alongside the impressive battery of Lancashire boilers and is now the centrepiece of the Cefn Coed Colliery Museum.

74

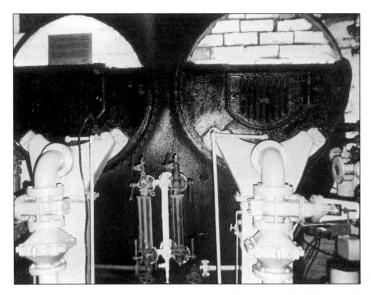

Cefn Coed Colliery boilers in 1968. After the closure of Cefn Coed Colliery a short-lived drift operated for a few years until the New Blaenant Drift Mine was commissioned to work the Rhondda No.2 seam. A 1,984ft-conveyor plane, at a gradient of 1 in 3.7, was driven. A modern manriding train carried 120 men into the mine, passing over the River Dulais in a covered bridge, which carried the outgoing conveyor. The mine had ten miles of underground roadways, with three miles of high-speed conveyors.

Left: Blaenant New Drift Mine in 1979. *Right*: Blaenant New Drift Mine eimco bucket and stage loader in 1979. Blaenant New Drift Mine was completed in 1976 at a cost of £1.4 million, together with general surface reorganisation, further streamlining this highly-productivity mine, was finished in 1983. In 1979 the colliery employed 640 miners and worked the Rhondda No.2 seam producing 512,000 tons, mostly for Aberthaw Power Station.

Blaenant Drift Mine in 1979. The estimated workable coal reserves were 9.2 million tons and with the productivity at this rate Blaenant was rightly regarded as a long-term (promises!) link in the chain of South Wales pits which represented the insurance for a fuel-hungry future! Despite the rich fault-free deposits the mine was closed on 11 May 1990 by British Coal (NCB).

Blaenant Colliery New Drift lamp check. In 1967 the Cefn Coed manager was H. Williams (5,526 First Class) and the undermanagers were E. Williams (6,143 Second Class) and D.R. Evans (6,999 First Class). In 1967 the Blaenant manager was T.J. Davies (4,077 First Class) and the undermanager was W. Harries (3,502 Second Class). Today at Cefn Coed Colliery Museum visitors (free entry) can discover the fascinating story of coalmining at what was once the deepest anthracite coalmine in the world. The underground gallery is accessible for all visitors. Cefn Coed Colliery was closed on 20 April 1968 by the NCB. New Blaenant Drift Mine was closed on 11 May 1990 by British Coal.

Treforgan Drift Mine, near Crynant, in 1970. Treforgan Drift Mine was sunk between 1963 and 1966, on the site of the old Llwyn Onn Colliery, which was abandoned in 1927. Treforgan was one of the most successful mines in the South Wales Coalfield. This highly-automated drift maintained a consistently high annual output of valuable anthracite coal for the domestic and power station markets and had a number of coalfield productivity records to its credit.

It was originally intended to be a short-life mine, but its impressive productivity, coupled with the constant need to meet the widespread demands for anthracite coal, led the NCB to take another look at the pit's future. Although it was surrounded by older closed collieries, Treforgan sat on much deeper reserves estimated at more than 20 million tons.

In 1976, as it became obvious that the life of the Red Vein seam was limited, the NCB decided to invest £7 million to enable Treforgan to open up new workings in the Nine-Feet and Bute seams, which were situated in the lower coal measures to the northeast.

Utilising the first 50yds or so of one of the old drifts, with its expensive equipment, the new mine then branched off in a different direction and brought its coal out from a depth of 2,300ft. Work became well under way on the new 870ft access drift and twin 1,500-yard roadways, which formed the skeleton of the new mine. On the surface, an entirely new layout supplemented the underground programme. In 1976, with a manpower of 484, it produced an annual saleable output of 168,000 tons; it produced an average weekly saleable output of 4,413 tons; the average output per man/shift at the coalface was 5 tons 1cwt; the average output per man/shift overall was 2 tons 4cwt; the deepest working level was 1,500ft; the number of coal faces working were two. The drift length was 779/848yds gradient 1 in 3; manriding capacity per train journey 94; coal conveying capacity per hour from pit 290 tons; type of coal was anthracite; the markets were power station/domestic; fan capacity was 174,000cu.ft per minute; average maximum demand of electrical power 4,434kW; total capital value of plant and machinery in use £1.1 million; estimate workable coal reserves 20 million tons. In 1967 the Treforgan manager was R. Walker (8,060 First Class) and the undermanager was B. Langdon (8,970 First Class). In 1973 over £100 million was earmarked for major modernisation projects for the South Wales Collieries and this mine was one of them.

Treforgan Drift Mine gleith obel rapid plough in 1970. A plough is a coal cutting machine used in longwall mining which planes a narrow strip of coal from 2in to 6in thick as it is hauled along the face in either direction, by an endless steel chain, allowing coal to fall onto the armoured flexible conveyor. The normal speed of a plough was 75ft per minute. It was developed in Germany in the 1940s and became popular in the South Wales Coalfield. The coal seams worked were: Red Vein of Abernant; Upper Six-Feet (Upper Black); Lower Six-Feet (Lower Black, Four-Feet or Cornish); Nine-Feet; Bute.

Treforgan Drift Mine manriding spake in 1976. The manriding capacity per train journey was ninety-four miners including the driver and guard. The driver is given directions in writing from the manager with respect to loads, speeds and all precautions necessary for the safe running of the locomotive. The spake travelled at a maximum speed of 3mph when carrying miners to their place of work. Treforgan Drift Mine was closed in 1985 by the NCB.

A Strike – A Leader – A Scab

In 1925, the coalowners decided they needed more wage cuts and once again the miners protested. The Government stepped in with subsidies to keep wages at their existing levels while a Royal Commission examined the industry. The Commission recommended in 1926 that the industry should be reorganised; some wage cuts would eventually be needed, it said. The owners demanded immediate cuts. Once again the miners refused and once again they found they had been locked out of their mines. Fellow workers decided to support the miners and a General Strike began on 3 May 1926. The strike lasted just nine days. The miners were eventually forced to accept the coalowners' terms and wage levels and returned to work on 1 December 1926.

One leading union leader during the strike was Arthur Horner (1894-1968). We may not agree with his politics, but his ability and his loyalty to his fellow men go without question. In May 1919 the authorising Council was asked to take steps to secure the release of Arthur Horner who was in prison as a conscientious objector. In his first meeting he repudiated the rumour of his not being willing to fight and stated that he was willing to shoulder a rifle to fight for the working classes, but not for the enemy of the workers – the capitalist, with unemployment so high and the means test resented. These were the words people wanted to hear – it was fighting talk, nectar to the ears. He saw representing the men in the workplace as quite separate from being a communist politician and showed it in negotiations on their behalf, which won him the respect of his work mates. His ability to negotiate from impossible positions made him stand out as a leader. Arthur Horner began to organise the education of workers; some areas became communist strongholds as awareness of 'people power' began to take hold. The means test was one of the most hated pieces of legislation as it preyed on the dignity of people who, with little money to spend, were forced to sell their furniture or pawn valuables before receiving 'Lloyd George' which was the nickname given to benefit at that time. He believed in the destruction of customs and price lists, the opposition to pit closures and the destruction of the company union, as priorities. This of course would lead to the lessening of hardship among people and was therefore out of step with the communist's way of thinking. Arthur Horner built up a loyalty with his followers that no other leader achieved and their dedication to him was to last throughout their lives. In 1936 Arthur Horner was elected the first President of the South Wales Miners Federation. During his Presidency an agreement was reached ending the company union and later he was instrumental in destroying the means test. In 1945 Arthur Horner played a leading part in the formation of the NUM and in 1946 became the first communist president of the union. He was involved in the discussions with the TUC and the Labour Government, which brought about the nationalisation of the mines and secured many important reforms, including new legislation for safety measures, holidays with pay, a retirement pension scheme and the raising of wages from among the lowest to among the highest.

Jack London's definition of a Scab:

> After God had finished the rattlesnake, the toad and vampire, He had some awful substance left with which he made a scab. A scab is a two-legged animal with a corkscrew soul, a waterlogged brain, a combination backbone of jelly and glue. Where others have a heart, he carries a tumour of rotten principles. When a scab comes down the street, men turn their backs; the angels weep in heaven and the Devil shuts the gates of Hell to keep him out. No man has a right to scab so long as there is a pool of water to drown his carcass in, or a rope long enough to hang his body with. Judas Iscariot was a gentleman compared with a scab. For after betraying his Master he had character enough to hang himself. A scab has not. Esau sold his birthright for a mess of pottage. Judas Iscariot sold his Saviour for 30 pieces of silver. Benedict Arnold sold his country for the promise of a commission in the British Army. The modern strike-breaker sells his birthright, his country, his wife, his children and his fellow men for an unfilled promise from his employer. Esau was a traitor to himself; Judas Iscariot was a traitor to his God; Benedict Arnold was a traitor to his country. A Strike-Breaker is a Traitor to his God, his country, his wife, his family and his class. A REAL MAN NEVER BECOMES A STRIKE-BREAKER...

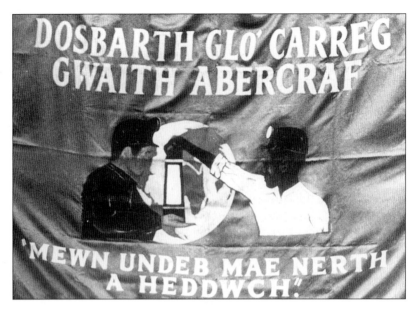

Abercrave Colliery, Abercrâf, Breconshire and Swansea Valley National Union of Mineworkers Lodge Banner, Abercraf Coal Section, Abercraf Works: 'In Unity There Is Strength And Peace'. In 1930 there were 140 abandoned mines listed in Breconshire. Abercrave Colliery was opened in 1892 by the Amalgamated Anthracite Collieries Ltd. The mine was producing coal at the northwestern boundary of the South Wales Coalfield in the Breconshire Coalfield, assumed by various authorities to have been approximately 74 square miles. Abercrave Colliery was closed on 17 March 1967 by the NCB.

Underground first aid container. The first aid container is a large canister hanging up in each underground district containing a stretcher, blankets, splints, dressings, morphia injection containers etc. Only the district deputy and medical room attendants are allowed to carry a morphia key and have the qualifications to administer morphia. The rescue station is a room on the surface of a mine, kept equipped for the use of rescue teams and officers, in the event of a fire, explosion etc. At every NCB colliery there was a trained and qualified rescue team from its own workforce, who were familiar with their mine.

Yniscedwyn (The Sced) Colliery, Ystradgynlais, Breconshire and Swansea Valley, beam engine winding house in 1900. Yniscedwyn Colliery was sunk in 1852. Close to the pithead were a variety of buildings associated with generating power. At the end of the eighteenth century and through much of the nineteenth century, tall beam engine houses housing mighty pumping engines with a myriad flimsy and variable wooden headframes attached dominated colliery sites. In the early nineteenth century the twin functions of pumping and winding were often combined. In 1967 the Yniscedwyn manager was D.L. Evans (4,140 First Class) and the undermanager was R.P. Herrington (6,084 First Class). Yniscedwyn Colliery was closed on 2 March 1968 by the NCB.

Bryn Henllys Slant Mine, Upper Cwmtwrch, Breconshire and Swansea Valley, gaffers in 1890. Bryn Henllys Slant Mine was opened in 1872. The slant had a water wheel to provide power for the mine. The mine was producing coal at the northwestern boundary just inside Breconshire in the South Wales Coalfield when acquired from the Anthracite Colliery Co. on Vesting Day in 1947 by the NCB. Bryn Henllys Slant was closed in August 1955 by the NCB.

Abernant Colliery, near Pontardawe, Swansea Valley, Glamorganshire, during reconstruction in 1956. Abernant Colliery was situated in a wide open valley which is bounded by the Gardener's Fault to the east and the Duffryn Fault to the west. The sinking of Abernant Colliery was commenced in 1954 and completed in 1958 at a cost of £10 million, to work the Red Vein at a depth of around 1,200ft. The shafts were also sunk down to a point just below the Peacock Vein, which is said to be the finest anthracite in the world. No.1 shaft had a depth of 2,510ft and the No.2 had a shaft depth of 2,961ft. In the early days of working the colliery encountered severe geological conditions, which were overcome.

Abernant Colliery in 1965. By the late 1970s the colliery take was eight square miles with forty-four miles of underground roadways and more than ten miles of high-speed belt conveyors. The colliery was one of the first mines to work on the retreat system of mining. Completed in 1958 at a cost of £10 million, 666 miners produced 182,000 tons of anthracite coal yearly, mostly from the Red Vein seam. New developments and hi-tech coalface costing £5 million was installed in 1986. Abernant Colliery was closed in 1988 by the NCB.

Clydach Merthyr Colliery, near Clydach, in 1918. Clydach Merthyr Colliery was opened in 1863 by the Graigola Merthyr Co. On 6 February 1884 the accident reports show that twenty-five-year-old collier Rees Bevan was killed by a fall of coal. In 1954 with a manpower of 523 it produced 115,571 tons and in 1956 with a manpower of 524 it produced 114,807 tons. Clydach Merthyr Colliery was closed in June 1961 by the NCB.

Scott's Colliery winding engine house, Llansamlet, in 1980. In 1770 Scotts Colliery was partly sunk by Capt. Scott, but due to financial problems the sinking came to a halt in 1771. The shaft was finally sunk in 1819. The Cornish pumping and winding engine house was built by coalowner John Smith between 1817 and 1819 to house a low pressure Boulton & Watt steam engine pump. The pit is the earliest winding engine house still existing that housed a rotary steam engine. Scotts Colliery was closed in 1838.

In 1700 Swansea was the largest coal port in Wales, in 1798 the Swansea to Abercrâf canal opened, in 1850 the railways came to Swansea and in 1913 there were forty-six collieries working in the district.

Coal Owners' Ten Commandments to Miners

1. Thou shalt have no other Master but me.

2. Thou shalt not make for thyself any comforts, nor the likeness of anything to thine own interest, neither on the earth, nor above, nor in the mine.

3. Thou shalt bow down to me, and worship me, for I am thy Master, and a Jealous Master; and I will show you no mercy, but endeavour to make you keep my Commandments.

4. Thou shalt not take the name of thy Master in vain, lest I sack thee at a minute's notice.

5. Honour thy Master, and his Stewards, and his Deputies, that thy days may be short, and few, for I shall not want thee when thou gettest old and not able to work; thou wilt have to end thy days in the poor-house.

6. Remember that thou workest six days with all thy might and with all thy strength, and do all I want thee, but the seventh day thou stoppeth at home, and do no manner of work, but thou shalt do all thou canst to recruit thine exhausted strength for my service on Monday morning.

7. Thou shalt have no other Union, as it is against my will.

8. Thou shalt always speak well of me, though I oppress thee; thou shalt be content if I find thee work and pay thee what I think well; thou shalt starve thyself and thy children if it is anything to my interest; thou must only think of me, not thyself.

9. Thou shalt have no meetings to consider thine own interest, as I want thee to keep thyself in ignorance and poverty all the days of thy life.

10. Thou shalt not covet thy Master's money, nor his comfort, nor his business, nor anything that is his; thou shalt not grumble at anything, as I want to reign over thee, tyrannise thee, and keep thee in bondage all the days of thy life.

Skewen Miner, c.1928

South Wales Areas in 1961 were: No.1 Swansea; No.2 Maesteg; No.3 Rhondda; No.4 Aberdare; No.5 Rhymney; No.6 Monmouthshire; No.7 Forest of Dean; No.8 Bristol and No.9 Neath. The chairman was A.H. Kellet; Swansea No.1 area general manager in 1961 was J.G. Tait; area production manager was J.D.H. Davies; group managers were: No.1 A. Thomas; No.2 J.H. Griffiths; No.3 D.W. Griffiths. The No.1 group of collieries were: Blaenhirwaun; Carway; Cynheidre; Crosshands; Great Mountain; Pentremawr. The total output was 465,801 tons; total manpower was 2,810. The No.2 Group of Collieries were: Garngoch; Morlais; Mountain; Brynlliw. Their total output was 435,477 tons and total manpower was 1,987. The No.3 group of collieries were: Clydach Merthyr; Daren; Felinfran; Graig Merthyr; Hendy Merthyr. Their total output was 478,452 tons and total manpower was 1,449.

Prior to reaching its destination, domestic or industry, minerals found associated with a coal seam must be extracted at the washery, such as:

Balls:	nodules of ironstone
Bast:	cannel or cannelly mudstone
Bastard:	applied usually to non-typical or intermediate lithologies, thus 'bastard clift', usually a strong sandy mudstone not sandy enough to be called rock
Brass:	hard bands, lenses, or nodules rich in pyrite; Clift (or cliff): blocky mudstone or silty mudstone
Clod:	soft mudstone forming a band in a coal seam, or that part of the roof which comes away with the coal during working
Dirt:	any band other than coal within a seam
Mine:	ironstone, mine ground is clift with much ironstone in bands or nodules
Rashes or Rashings:	either soft carbonaceous shale with streaks of coal or highly disturbed, slickensided, comminuted shale or mudstone formed by movement parallel to the bedding and usually associated with the roof or dirt bands in coal seams
Rock:	hard massive sandstone
Stone:	a hard band usually within a coal seam.

Left: Gwaun Cae Gurwen Colliery, Maerdy Pit, Gwaun Cae Gurwen, Aman Valley, in 1915. The colliery consisted of four shafts which were: Maerdy, East, Steer and Buckland. Gwaun Cae Gurwen Maerdy Pit was opened in 1886. Cae Gurwen Buckland Pit was 600yds in depth and opened in 1927, but was never used. *Right*: Gwaun Cae Gurwen Colliery, East Pit underground jigger and gate belt conveyors in 1953.

Gwaun Cae Gurwen Colliery, Maerdy Pit, in 1919. On 8 October 1889 the accident reports show that thirty-four-year-old collier John D. Davies was killed by a fall of clod 12in thick from between two slips in the roof, forming an acute angle with the coalface. There were props on each side, but none under the part which fell. He was working in the Big Vein seam, which had a 4ft high section of coal.

Gwaun Cae Gurwen Colliery, East Pit winding engine house in 1919. Gwaun Cae Gurwen East Pit was opened in 1910. On 30 November 1889 at Gwaun Cae Gurwen Maerdy Pit, the accident reports show that seventeen-year-old collier David Thomas was killed by a fall of cliff (clift or cliff is blocky mudstone or silty mudstone) 7ft x 4ft x 2ft to 0in, part of the rippings, which had one prop near the middle. The cliff fell from a flat slip. The seam had a 3ft 6in high section of coal.

Gwaun Cae Gurwen Colliery Steer Pit in 1959. Gwaun Cae Gurwen Steer Pit was opened in 1922. On 6 March 1890 at Gwaun Cae Gurwen Maerdy Pit, the accident reports show that sixty-five-year-old labourer William Williams was killed when he was squeezed between a dram and the side of a slope, and on 24 April 1912 they show that thirty-two-year-old collier J. Gourmill was fatally injured.

Gwaun Cae Gurwen Steer Pit, Gen. Bernard Law Montgomery (Monty) on tour in 1947. Gwaun Cae Gurwen Maerdy Pit was closed in 1948, Steer Pit on 2 February 1959 and the East Pit in 1962 by the NCB.

Gelli Ceidrim No.1 Slant Mine, near Glanaman, Carmarthenshire and Aman Valley, in 1925. Gelli Ceidrim No.1 Slant was opened in 1891 by the Gelli Ceidrim Collieries Co. In 1954 with a manpower of 229 it produced 29,617 tons; in 1955 with a manpower of 257 it produced 30,811 tons and in 1956 with a manpower of 226 it produced 26,608 tons. The local mines supplied the nearby Garnant Plate Works. Gelli Ceidrim No.1 Slant Mine was closed in November 1957 by the NCB.

Top: A wooden sledgehammer c.1750. The sledgehammer was found near Ammanford, Carmarthenshire, in the Stanllyd seam during opencast work in 1986. Opencast mining began on an extensive scale during the Second World War. For several years half of Britain's anthracite coal output came from opencast, all of it from the South Wales Coalfield. Opencast high quality fuels helped to improve blends from deep mines in the region and assisted in making them viable. British Coal's (NCB) opencast operation in South Wales had been a stable source of production in the region. It has produced over 70 million tons of fuel, all of it won profitably.

Bottom: Saron Colliery, near Ammanford, cutting the first sod in 1915. Saron Colliery was opened in 1915 by Blaina Colliery Co. In 1947 the colliery employed 361 miners. The seams worked were the Stanllyd, Lower-Triquart and the Lower-Pumpquart. Saron Colliery was closed in September 1956 by the NCB.

Ammanford Slant Mine in 1975. Ammanford 1 in 4 Slant was opened in 1891 by Ammanford Colliery Co. In 1921 the colliery employed 573 miners. In 1924 it was owned by United Anthracite Collieries Ltd. In 1927 it was owned by Amalgamated Anthracite Collieries Ltd, then the largest combine. It was replaced by Betws New Mine. Ammanford Slant Mine was closed in 1975 by the NCB.

A member of the mines rescue team in the process of a rescue attempt in the early 1930s. In 1886 a Royal Commission recommended the establishment of mines rescue stations; Crumlin and Aberaman were established in 1908, New Tredegar 1910. They did not become widespread until the Coal Mines Act of 1911 made them compulsory. Following the decline in coal mining activity throughout Britain, the Dinas Mines Rescue Station has now become the only operational centre in the South Wales Coalfield. The rescue station continues to serve Tower, Betws, Aberpergwm, Pentreclwydau South Mine, five small mines, the Forest of Dean and the West Country stone mines. The station duty room is manned daily and is able to turn out immediately, with at least one team of rescue men fully equipped with a rescue officer. The emergency tools that are used by the mines rescue team at an underground rescue are made of nonferrous (non-sparking) metal.

Betws Drift Mine, near Ammanford, in 1980. Betws Drift Mine was officially opened on St David's Day 1978 by HRH the Prince of Wales and began production as the first new mine in South Wales for more than ten years after driving twin access tunnels of 2,263yds into the Red Vein anthracite seam; further extensions started in 1985 costing £15 million including a new drift and a wide-diameter ventilation shaft drilled from the surface.

Betws Drift Mine, roof bolting in 1986. Roof bolting is a method of stabilizing a roof by drilling deep holes and fixing bolts to which are attached wire mesh panels. Where a coal seam roof is unstable, long wooden or plastic bolts can be bonded into place to give temporary support; when the coal is cut these bolts cause no damage to the coal cutting machinery. Steel bolts are used in roadways but normally steel arches (rings) are used. Betws Drift Mine was started in August 1974 and when opened in 1978, it was one of the most modern and comprehensively equipped mines in western Europe. In 1980 Betws Drift Mine received the premier Business & Industry Award in recognition of human and social responsibility and environmental quality; 736 miners produced over 369,075 tons. The mine was closed in June 1993 by British Coal. Betws Anthracite Ltd Drift Mine reopened in May 1994 with a management buy-out supported by a NUM workforce at the mine and is coaling (producing coal) today May 2002.

Pantyffynnon Upcast Slant Mine, Pantyffynnon, near Ammanford, in 1965. Pantyffynnon anthracite mine was opened in 1902. The seams worked were the Brynlloi, Lower-Pumpquart, Stanllyd, Upper-Pumpquart and Rock Vein. The Upcast Slant was a secondary slant that returns stale air to the surface. The exhausting fan to extract this air is situated in the background of the photograph.

Pantyffynnon Slant Mine, near Ammanford, in 1960. Two colliers hewing coal by hand at the coalface in the Stanllyd seam in 1960. On 18 January 1906 the accident reports show that eighteen-year-old pumpman Griffith Davies was killed; on 7 February 1907 they show that twenty-five-year-old signaller John Oliver Thomas was killed. In 1913 the colliery employed 138 miners; in 1947 the manpower was 458 and in 1954, with a manpower of 491, it produced 89,000 tons from the Stanllyd seam. The mine was producing coal when acquired from Anthracite Collieries Ltd on Vesting Day 1947 by the NCB. In 1967 the Pantyffynnon manager was H.E.G. Roberts (5,889 First Class) and the undermanagers were H.B. Davies (5,478 First Class) and L. Lounds (5,689 First Class). Pantyffynnon Slant Mine was closed on 31 January 1969 by the NCB.

Wernos Drift Mine, near Pantyffynnon, in 1964. Wernos Drift Mine was opened in 1904 by Rhos Colliery Co. Ltd; other owners include the United Anthracite Collieries Ltd. The mine was producing coal when acquired on Vesting Day 1947 by the NCB and employed forty-one miners. In 1961 with a manpower of 549 it produced 97,279 tons. Wernos Drift Mine was closed in November 1965 by the NCB.

Underground wooden pit props and flats in the supply road in 1964. The erection of a wooden prop, also called a post, supported the roof and slips at a coalface. A flat is a wooden post, cut lengthways, placed beneath the roof in a face and secured there by two posts, supporting that area of the face and set at measured distances, as stated by the support rules of the district. By the 1960s hydraulic props were being manually pumped against roof steel bars, with a key (a lever type of tool). It was a safer and easier method of extraction.

Graig Merthyr Drift Mine, near Pontardulais, Glamorganshire, in 1918. Graig Merthyr Drift Mine was opened in 1873 by Graigola Merthyr Co. On 3 July 1873 forty-year-old collier John Jones was killed by a fall of coal and on 15 January 1874 twenty-five-year-old collier D.L. Evans was killed by a fall of coal. The seams worked were the Graigola, Swansea Six-Feet and the Three-Feet.

Graig Merthyr Drift Mine in 1965. Miners returning to the surface following a day's work underground. In 1954 with a manpower of 651 it produced 222,846 tons and in 1956 with a manpower of 651 it produced 229,612 tons. In 1967 the Graig Merthyr manager was T.E. Banks (4,681 First Class) and the undermanagers were A. Jones (7,523 First Class) and T.E. Lloyd (3,335 First Class). Graig Merthyr Colliery was closed on 23 June 1978 by the NCB.

Left: Brynlliw Colliery, Grovesend, Upcast Shaft in 1966. Brynlliw Colliery was opened in 1908, ceased production in 1925 and was abandoned in 1927. The NCB re-opened the colliery in 1954 and linked it to Morlais Colliery, Llangennech, Carmarthenshire. *Right*: Brynlliw Colliery underground Hunslet diesel locomotive in 1976. Brynlliw Colliery was originally sunk between 1903 and 1908 by Thomas Williams (Llangennech) Ltd, to work dry steam coals from the Swansea Four-Feet seam at a depth of around 340 yards. In 1914, the deeper and slightly thicker Five-Feet seam was also developed, but in 1925 the pit became a victim of the prevailing depression and was closed in 1927.

It remained unproductive until 1954, though the shafts and headgear were retained. In that year, the NCB approved a major redevelopment, aimed at restoring Brynlliw to a working mine, at a cost of £4.8 million. The project included the repair and deepening of the two shafts; the driving of several thousand yards of new underground roadways for loco haulage; new headgear, equipped for skip (a container that carried coal through the shaft) winding of coal in the No.2 shaft and for manriding and materials in the No.1 shaft; a modern washery, capable of handling 270 tons an hour run-of-mine coal and the construction of new pithead baths and administrative buildings. By 1961, the colliery was back in production and by 1968 was showing an annual output of well over 300,000 tons, from the Swansea Three-Feet and Six-Feet seams.

Four and a half miles to the east, across the Loughor Estuary, the smaller Morlais mine was sunk in the early 1890s, also by Thomas Williams, on the site of an earlier slant. Students of local history perhaps best know it as the last operational pit in the coalfield to use steam powered winders. The 13ft diameter partly unlined shaft was used for winding men and materials.

In 1976 a 1,013ft underground roadway was driven from the Morlais workings, at a gradient of 1 in 3.5, passing directly under the estuary to link up with Brynlliw, forming a single, highly efficient production unit. From mid 1977 until its closure, all the Morlais coal was transported underground and wound at Brynlliw for washing, eliminating considerable surface transport. The conveyor system incorporates a 200-ton 'buffer' bunker, to maintain the flow of coal in the event of winding difficulties at Brynlliw. Production at the mine in 1976 was concentrated in the Three-Feet and Six-Feet seams at Brynlliw and in the Five-Feet at the Morlais end and a combined workforce of 836 miners produced an average 6,000 tons a week.

94

Brynlliw Colliery Löbbehobel plough in 1955. The twelve square miles of workings were affected by six major faults, which displaced the coal seams by between 150ft and 1,000ft. In the Six-Feet, roof conditions were notoriously difficult and face lengths generally had to be curtailed to a maximum of 375ft, considerably shorter than the coalfield average.

Brynlliw Colliery underground conveyors in 1955. In 1976 within the mining programme, there were 13.5 miles of underground roadways, carrying almost four miles of high-speed belt conveyors. The mine produced dry steam coal primarily for blending into the special mixes that were used at power stations, with smaller quantities going to the domestic market. The seams worked were the Swansea Four-Feet, Swansea Five-Feet, Swansea Six-Feet, Swansea Three-Feet, Swansea Two-Feet and the Hughes. Brynlliw Colliery was closed in 1985 by the NCB.

Morlais Colliery, Llangennech, winding engine in 1980. Morlais Colliery was opened in 1894. The mine was producing coal when acquired on Vesting Day 1947 by the NCB and employed 303 miners. The seams worked were the Six-Feet and the Four-Feet. In 1967 the Morlais manager was C. Bryant (4,649 First Class) and the undermanager was W.J. Challis (7,734 First Class). Morlais Colliery was closed on 8 May 1981 by the NCB.

Broad Oak Colliery, Loughor, winding engine house and double beam winding engine in 1944. This mine was situated on the banks of the River Loughor and was opened in 1880 by Broad Oak Colliery Co. A beam engine is an atmospheric or steam engine in which the piston rod of the vertical cylinder is attached to one end of a massive, centrally pivoted beam, the other end of which is attached to the plunger rod for operating a pump, or, in a rotative beam engine, the crank of a flywheel. In the early nineteenth century it was commonly used for pumping mine water, and winding coal and minerals etc. Broad Oak Colliery was closed in March 1948 by the NCB.

Caenewydd Drift Mine also known as the Mountain Drift, Gorseinon, Lliw Valley, in 1895. The miners are, left to right, back row: William 'Billy Sais' Grenfell, David Rhys Grenfell, David 'Daffyd Banc-Yr-Eithin' Rees, Henry Howard, William 'Billy Boy' Evans. Front row: Henry 'Rocking' Thomas, William John 'W.J.' Grenfell, David 'Coalbrook' Mathews, Thomas 'Twm Arall' Evans. The mine was owned by the Caenewydd Colliery Co. Ltd in the 1920s. Caenewydd Drift Mine was producing coal when acquired on Vesting Day 1947 by the NCB.

Artist Ken Williams' 'Impression of the Underground Chapel' at Mynydd Newydd Colliery, Fforest Fach, near Swansea. The pit was sunk in 1843. An explosion on 23 March 1869 caused the death of three miners and as a result of this tragedy the men decided to establish a chapel underground in the Six-Feet seam lying at a depth of 774ft. There, every Monday at 6:00 a.m. for the next sixty-three years until 1932, a service was held before the men proceeded to their working places. Mynydd Newydd Colliery was closed on 1 July 1955.

In everlasting memory to the miners who lost their lives in the Dulais and Swansea Valleys, and the Breconshire, Aman and Loughor Valleys.

Date	Mine	Lives Lost
21 July 1852	Mynydd Bach Glo	Gatekeeper Edward David (12) killed when he was kicked by an earth horse.
11 February 1853	Aman	Labourer John Davies (35) killed when he was struck by a handle of a windlass.
11 February 1853	Tryrcenol	Collier John Mathews killed by a fall of stone from the roof.
10 August 1853	Yniscedwyn	Labourer Richard Davies (11) killed when railway drams passed over him.
2 November 1853	Cwmfelin	Sarah Morris (14) killed when she was dazzled by a pit lamp and fell down the shaft.
14 October 1858	Primrose	Fourteen miners and seven horses killed by choke damp.
23 July 1870	Llansamlet	Nineteen miners killed by an explosion.
26 November 1873	Cae Cam	Sinkers T. Jones (42) and W. Davies (24) killed by falling down the shaft.
29 May 1874	Hafod Brynaman	Collier David Thomas (45) hurt by a fall of coal and died on 7 June 1874.
24 July 1874	Charles	Nineteen miners killed by an explosion.
24 August 1874	Abercrave	Sinker D. Handry (13) killed by an explosion.
17 October 1874	Hendre Ladis	Colliers S. Jones and H. Watkin killed by gunpowder igniting whilst the deceased and two others were ramming a hole.
20 November 1874	Brynhurgan	Overman W. Andrews (33) killed by falling down the sinking shaft.
31 March 1875	Ynysmedu	Collier David James (24) killed by a fall of stone.
30 April 1875	Wernddu	Collier T.W. Gibbs (35) killed by falling off a dram whilst riding down an incline, which was against the rules.
10 May 1875	Onllwyn	Collier John Morgan killed by a fall of coal.
10 May 1875	Seven Sisters	Sinker E. Morris (35) killed by falling down the shaft.
19 February 1876	Abercrave	Doorboy Owen Davis (13) killed by a fall of stone.
19 October 1876	Seven Sisters	Collier T. Lewis (19) killed by going past a danger signal into gas. Deceased left his safety lamp behind and went into gas and was suffocated.
8 March 1877	Worcester	Eighteen miners killed by an explosion.
8 March 1877	Wigfach	Nineteen miners killed by an explosion.
13 December 1884	Waunclawdd	Collier T. Williams (60) killed by a fall of roof.
24 August 1892	Hendre Ladis	Labourer John Davies (35) killed when he was struck by a handle of a windlass.
24 August 1892	Hendre Ladis	Six miners killed by a runaway spake.
27 January 1905	Elba	Ten miners killed by an explosion.

Two
Carmarthenshire in the South Wales Coalfield

The Carmarthenshire Coalfield is assumed by various authorities to have been approximately 228 square miles. Many communities were almost semi-rural and quite isolated. This was particularly so in the western part of the South Wales Coalfield. In these anthracite valleys the coal industry developed at a much slower pace and as a result the collieries and communities of west Wales were smaller than the rest of the coalfield and this was an area where the Welsh language remained strong and ties with local agriculture were so close that there were men who spread their work between being a miner and working on their own small holding or labouring on a farm. Together these differences resulted in a quite different way of life from the rest of the South Wales Valleys.

On the whole, the worldly lot of the anthracite miner appeared to be better off than that of the steam coal miner. Most of the mining villages were smaller and the cottages not huddled into untidy streets as they were in the central valleys. There is a great difference between these valleys, with their collieries in the main streets, and the clean open villages of Carmarthenshire. East, centre or west of the South Wales Coalfield one thing is certain: the coal industry dominated the life of the people of the South Wales Valleys and Vales.

The South Wales Collieries produced coal just under the sod from opencast, drifts, levels and slants, and from deep mines to obtain the lower coal measures deep down under the mountainous valleys and vales. In general, the upper seams are more bituminous than the lower, but the most marked change is, of course, when going from east to west, the coal being bituminous in the east and then passing through steam coal in the central valleys to genuine anthracite in the west. The exception lies to the south of the anticlinal line running east and west from Risca through Pontypridd to Port Talbot, where the coal is bituminous. With regard to the formation of anthracite, many theories have been made. It would seem, however, that the old theory that the anthracite area was depressed lower than the bituminous area, and therefore was subjected to greater heat, is untenable for several reasons. The majority of modern authorities incline to the opinion that it is due to original differences in the deposits. In addition, the anthracite seams were subjected to great heat and pressure as the result of special decomposition of the original mother substance. The character of the coal in the South Wales Coalfield can be approximately classified as: bituminous coal, thirty per cent; steam coal, forty-eight per cent; and anthracite coal, twenty-two per cent. This coal made the South Wales Collieries and South Wales world famous. In burning, it produced little smoke and the higher quality coal that was on the Admiralty lists were so free in this respect that it was universally known as Welsh Smokeless Steam Coal.

The work of miners changed dramatically with the wide scale introduction of coal cutting machines into the pits and particularly in the 'super pit', Cynheidre Colliery which stood on the flanks of the Gwendraeth Fawr Valley some four miles from Llanelli.

Cwmgwili Drift Mine and Spake, Cwmgwili, near Cross Hands, Carmarthenshire, in 1969. Just before the closure of Cross Hands Colliery in 1962 twin drifts Cwmgwili and Lindsay were driven into the Big Vein seam. In 1970 a £150,000 streamlining scheme merged the two mines and, despite severe geological conditions, it produced an extremely high tonnage of anthracite coal, making regular appearances in the productivity record tables with a work force of only 379 miners. In 1978 the annual output was 126,000 tons of anthracite from the Big Seam and the Hwch.

Cwmgwili Drift Mine power cutter and loader in 1972. This colliery was one of the few mines to operate a system of production headings or stalls using Joy loaders (cutting a maximum 2 tons of coal per minute) into the 1980s. Drift length 2,067ft; gradient 1 in 3.1; coal conveying capacity per hour from pit 450 tons; type of coal anthracite; markets domestic central heating and export; total capital value of plant and machinery in use £44,200.

Cwmgwili Drift Mine and Spake in 1976. All the coal was taken out the Lindsay end, direct on to road transport for treatment at the nearby Wernos Washery. The mine appears in the British Coal list of colliery closures as having closed in November 1980, but the mine was working under the ownership of Coal Investments plc until a management buy-out in 1994. The mine appears to have worked little or no coal after 1994 and was certainly closed in 1995.

New Cross Hands Colliery, Gwendraeth Fawr, in 1911. New Cross Hands Colliery was opened by the New Cross Hands Colliery Co. Ltd; it was also owned by Cleeve's Western Valleys Anthracite Collieries Ltd and Amalgamated Anthracite Collieries. The mine was producing coal when acquired on Vesting Day 1947 by the NCB and employed 601 miners. In 1960 with a manpower of 378 it produced 53,371 tons. New Cross Hands Colliery was closed in May 1962 by the NCB.

Dynant Fâch Colliery Co. Ltd, Bethesda Road, Tumble, Gwendraeth Fawr, in 2001. South Wales Collieries name signs along with the collieries have now become almost extinct. I received a friendly welcome by the miners on my visit to the mine in 2001. Dynant Fâch Colliery was opened in the early nineteenth century and over the years several drifts have been worked on this site. Dynant Fâch Colliery is now owned by Dynant Fâch Colliery Co. Ltd.

The direct (or main) hauling engine that was used during coaling in 1998. It was an electric single drum main rope haulage and the underground rail tracks were set at a 2ft wide gauge. When I visited the colliery in 2001 the preparation for the opencast method of mining was underway to exploit what is now becoming a rare and precious mineral called coal, Black Gold – Aur Du. No.1 Drift Mine was abandoned in 1995 and No.2 Drift Mine was abandoned on 18 December 1998. Dynant Fâch Colliery has a current operating licence which was granted on 22 December 2000.

Glyn-Y-Hebog Drift Mine, near Pontyberem, Gwendraeth Fawr, in 1960. The Glyn-Yr-Hebog Drift Mine was opened in 1892. On 1 October 1897 the accident reports show that an explosion injured one miner; on 28 May 1900 they show that an explosion killed two miners; and on 15 June 1914 a fireman was killed by suffocation from gas. Glyn-Yr-Hebog Drift Mine was abandoned in 1915.

No.3 Pentremawr Slant Mine, Pontyberem, Gwendraeth Fawr, in 1972. In 1870 four slants were driven near the anthracite outcrop, major faults were struck and the mine was abandoned. The mine reopened in 1895. In 1913 the Pumpquart Slant was opened. In 1939 the Big Vein was struck and proved very successful. The owners were the Pentremawr Colliery Co., Amalgamated Anthracite Collieries.

No.3 Pentremawr Slant Mine, Gwendraeth Fawr, gaffers and miners in 1912. In 1947 the colliery employed 718 miners; in 1960 with a manpower of 834 it produced 215,535 tons and in 1961 with a manpower of 846 it produced 188,736 tons. In 1963, mining operations at Cynheidre Colliery were extended to incorporate the Pentremawr Drift Mine. In 1967 the Pentremawr manager was V.K. Jones (5,660 First Class) and the undermanagers were D.J. Butler (5,943 Second Class) and R. Muir (5,287 First Class). The seams worked were the Glas, Big Vein and Pumpquart (Fivequart).

Pentremawr Slant Mine Spake in 1968. Anthracite was used for gas producers, malting, baking, horticultural work and cement manufacture. It has also been a large market for domestic heating, where it is renowned for its economy and slow burning qualities. Large coal was passed from the screens and conveyed into breaking machines, in which it was broken up into smaller pieces, the broken coal was then screened and graded into various sizes to meet the requirements of different markets. Pentremawr Slant Mine was closed on 31 March 1974 by the NCB.

Two miners boring the coal at the coalface in the early 1960s. There were various types of boring machines and drills used in the mining industry. In the early days of mining a sharpened, pointed iron bar was turned by a miner, while another struck it with a sledgehammer. Then from the 1850s came the breaster, a manual type again – one man with an iron cross against his chest and a handle between the cross and the iron drill; body pressure was applied while turning the handle.

A claw coal boring bit in the early 1960s. Prior to cutting the coal in poor roof conditions, a hole near the roof would be bored with a claw bit and a steel H-Bar inserted into the hole; the bar would then be propped against the roof and this would help to control the poor roof conditions. When I worked underground I used the compressed air (blast) stone boring machine, the blast and electric coal boring machine. An easy life compared to the old hand boring machines.

A colliery newspaper cartoon in 1947. 'Well I wanted to try the Coal recipe, but you would have your mother's. The NCB Coal magazines and the *Coal News* were popular with the miners and often available and read in the colliery canteen. In 1947 the *Coal* newspaper was 4*d* (approximately 2p). The cartoon shows another alternative use of a coal-boring machine and drill. One wonders what type of bit the miner was using?

A Jim Crow in 1970. The Jim Crow was used underground to straighten damaged rails following an accidental derailment of drams, and to bend and shape rails in the building of double partings and turns, etc., by the block layer (road man) who was skilled in laying and maintaining rail track throughout the pit. A double parting is a roadway containing one dramway which enters a section of a wider roadway containing two sets of dramway. It is a transfer area where a full journey of coal is deposited and another journey of empty drams is ready to be taken to the coalface.

Great Mountain No.1 Slant Mine, Tumble, Gwendraeth Fawr, in 1955. The Great Mountain No.1 Slant Mine worked the Four-Feet/Upper Six Feet (Big Vein) and the Lower Gellideg (Pumpquart) seams. Great Mountain No.1 Slant Mine was closed in 1962 by the NCB.

Great Mountain Colliery locally known as Tumble Colliery, in 1912. Great Mountain Colliery was opened in 1887 by Great Mountain Colliery Co. Ltd and was later owned by Amalgamated Anthracite Collieries Ltd. The seams worked were the Big Vein, Green Vein, Braslyd, Grad, Stanllyd and Triquart. In 1955 with a manpower of 900 it produced 172,437 tons. No.3 shaft depth was 2,286ft with a diameter of 18ft and became Cynheidre No.3 shaft; Cynheidre was situated two and a half miles away. Great Mountain No.1 and No.2 Colliery was closed on 5 May 1962 by the NCB.

Left: Colliery craftsmen Phillip Quick and Trevor Jones taking a short break in 1953. *Right*: Miners Federation Of Great Britain, National Union Of Mineworkers drinking mug: 'The Past We Inherit, The Future We Build'. The Fed was the popular name for the South Wales Miners Federation. Throughout November 1971, a rash of unofficial strikes over pay disputes caused great unrest in the Welsh mining communities. This industrial action brought matters to a head and a strike was called on 9 January 1972. The national strike, the first since 1926, resulted in the whole of the South Wales Coalfield being brought to a standstill. It was to be almost two months before coal was again raised, but the dispute, which had a devastating effect on British industry, saw the miners return to work as victors. To some, it was in some small way a vindication of their fathers and grandfathers who suffered such an humiliating defeat forty-six years earlier. The strike had shown that despite the increased use of oil and nuclear power as alternative energy sources, the nation's prosperity still relied heavily on coal. A further strike in 1974 again saw the union locked in a dispute, which ultimately brought down the Heath Government.

For many observers, the NUM's response to the NCB's announcement on 6 March 1984 of a cut in national capacity of 8 million tons of coal per annum and, more immediately, of the imminent closure of the Cortonwood Colliery in Yorkshire, merely confirmed a stereotype of the miners' historical role in opposing Conservative Governments. Within a fortnight every South Wales miner was on strike. Six Liverpool gravediggers, members of NUPE, organised a collection for the Miners' Solidarity Fund. They have also offered their services to Ian MacGregor – free of charge. A single miner's lamp raised £2,000 for the strike fund.

On St David's Day, 1 March 1985, the South Wales A0rea Conference took the momentous decision, by an overwhelming majority, to call for a national return to work without an agreement. For the mining industry, the 1984-1985 strike was a fight to save jobs. However, perhaps many more view the loss of jobs as a small price to pay for an end to the terrible toll of human life, the suffering and the desecration of a once beautiful landscape, which were hallmarks of an era when coal was king. There was always tremendous courage and staunch camaraderie with the South Wales miners in the deep, fiery and dangerous pits.

Cynheidre Colliery, near Llanelli, in 1979. Cynheidre Colliery stood on the flanks of the Gwendraeth Fawr Valley four miles from Llanelli. In a coalfield notorious for its geological problems, Cynheidre fought harder than most for its valuable anthracite supplies. In 1963, operations were extended to incorporate the nearby Pentremawr Colliery, based on a series of slants dating back to 1895. Earlier slants had been driven at Pentremawr in 1870 but had struck major faults and were abandoned.

The NCB had estimated that the colliery would have a workable reserve of 60 million tons of high-grade anthracite, but the colliery was sunk in the most geologically disturbed conditions in the South Wales Coalfield. Major faults lie between the No.1 and No.2 shafts and the No.3 shaft at Pontyberem two and a half miles away, including the formidable Lletty Wilws which involves a drop of 240ft in the same seam from one side to the other. The most geologically disturbed conditions in South Wales had constantly inhibited operations and restricted the production of valuable anthracite from the 'super pit'. Although as many as eight coal seams have been worked during the life of Cynheidre, the operations in 1976 took coal entirely from the Big Seam at a depth of 1,980ft. Newer measures were being taken at the pit that included a project driving into the Pumpquart seam, which lay 218yds below. The £3.2 million development exposed further coal reserves, which were estimated at more than 12 million tons. The coalfaces boosted output considerably, though, as always, geological conditions played their part in the extraction of this valuable commodity. The total working area of the colliery lay beneath 228 square miles and involved more than fifteen miles of underground roadways although around forty miles have been driven during the life of the mine. Within the network there were more than four miles of high-speed belt conveyors in daily use.

Exceptional geological pressures prevalent in the area made Cynheidre particularly prone to the problem of 'outbursts'. Gas and dust bursting forth under pressure is an obvious hazard below ground and close scrutiny had to be given to this possibility at every stage in the planning of new coalface development. Sixty-seven outbursts were reported between October 1962 and 1971, but fortunately none resulted in any loss of life until 6 April 1971, when a major outburst of coal and firedamp killed six men with sixty-nine others suffering varying degrees of asphyxia. At the time of the outburst the combined colliery produced about 20,000 tons of coal annually and employed 1,178 men underground and 422 on the surface.

Cynheidre Colliery during reconstruction in 1961. A bowk (bowc) is a bucket or vessel and was used for transporting men and materials; a flat temporary winding rope was also used during shaft sinking. As well as geological pressures underground water also added to the problems of the mining operation and 600hp pumps removed between ten and fifteen million gallons a week from the workings. As might be expected in such a modern mine, up to the minute equipment was used wherever possible. In 1976 with a manpower of 1,224 it produced an annual saleable output of 158,000 tons; it produced an average weekly saleable output of 6,000 tons; average output per man/shift at the coalface 6 tons 17cwt; average output per man/shift overall 1 ton 5cwt; deepest working level 1,978ft; number of coal faces currently working were two; No.1 shaft depth 2,395ft; diameter 24ft; No.2 shaft depth 2,368ft; diameter 20ft; No.3 shaft depth 2,286ft; diameter 18ft. No.4 shaft depth 1,986ft; diameter 18ft; coal winding capacity per cage wind 18 tons; manwinding capacity per cage wind 100; winding engines horsepower 850/2,700; average washery throughput 350 tons per hour; type of coal anthracite; markets domestic; total capital value of plant and machinery in use £812,640 and the estimated workable coal reserves were 12 million tons. Coal Seams: Four-Feet/Upper Six Feet (Big Seam); Lower Six-Feet (Green Seam); Red Seam (Ddugaled Seam); Lower Red (Hwch Seam); Upper Nine-Feet (Stanllyd Seam); Lower Nine-Feet (Gras Uchaf Seam); Bute (Brasslyd Fawr); Yard, (Gwendraeth Seam); Lower Gellideg (Pumpquart Seam).

An NCB press release dated 26 February 1986 stated 'that the board had approved investment of £30 million to mine rich new reserves of anthracite to secure mining jobs at Cynheidre Colliery well into the next century'. With the new Carway Fawr development, the investment in South Wales stood at £80 million that year. In spite of this massive investment Cynheidre Colliery was closed in January 1989 by British Coal (NCB).

The area has since been cleared and landscaped although a few railway sidings remain. The colliery was on the route of a historic railway line – the Llanelli & Mynydd Mawr Railway which when it began traffic in May 1803 as the Carmarthenshire Tramway or Railway was Britain's first operating public railway. A new charitable company formed in 1999, the Llanelli & Mynydd Mawr Railway Co. Ltd, is well advanced with its preparations for establishing an operating heritage railway and industrial heritage centre on the former Cynheidre Colliery site.

Carway Fawr Drift Mine, near Pontyates, Gwendraeth Fawr, in 1986. Mining in this area dates back to 1863. A surveyor checks the new development of the £30 million Drift Mine under construction near Cynheidre Colliery to mine rich new reserves of anthracite. Carway Fawr Drift Mine was abandoned on 13 November 1992 by British Coal.

Plas Bach Colliery, Pontyates, Gwendraeth Fawr, in 1896. The wooden headgear on the left in the photograph was the downcast shaft and the wooden headgear on the right was the upcast shaft. The downcast shaft was a ventilation shaft (drift, level or slant) where fresh air is drawn, or forced, into the workings. The upcast shaft was a secondary shaft that returns stale air to the surface. It normally contained an exhausting fan to extract this air. Plas Bach Colliery was abandoned in 1929.

Pant Hywel Level, near Llanelli, Gwendraeth Fawr, in 1970. The unusual entrance to Pant Hywel Level was by means of a cast iron dram bridge that was made by Waddle Lanmer in 1845. Pant Hywel Level was opened in 1845 by David Jones. In 1913 the mine employed thirteen miners. Pant Hywel Level was abandoned on 31 March 1925.

Pencoed Drift Mine, near Llanelli, Gwendraeth Fawr, in 1938. Pencoed Drift Mine was opened in 1867 by Syms, Williams & Co. In 1878 it was owned by Neville & Druce. In these early days coal was lying at the very root of our manufacturing prosperity and gave the country a position in the metallurgical trades which was second in the world only to the USA. Pencoed Drift Mine was abandoned in May 1981.

Conway Drift Mine, near Kidwelly (Joint River Beds) Gwndraeth Fawr and Fach Valleys, in 1955. Many of the coal seams in the area that were worked had local names and there has been considerable discussion as to the correlation of some of them, more especially in the case of the anthracite coal measures, owing to the disturbed nature of the ground. Names of the seams that were generally worked in the area were the Coed-Rhyal, Carway, Big, Green, Ddugaled, Hwch, Stanllydd, Gras Uchaf, Brasslyd, Gwendraeth, Triquart and Pumpquart. The Conway Drift Mine was closed on 1 July 1960 by the NCB.

Glyn Mawr Drift Mine, near Carmarthen, Gwendraeth Valley, in the 1950s. Glyn Mawr Drift Mine was opened in the 1880s and abandoned in the 1950s. Mining sites have been cleared and landscaped. However, throughout the region less spectacular evidence of coal mining can be seen in the form of collapsed level mouths, grassed-over spoil heaps or small groups of tumbledown buildings. Carmarthenshire has numerous beauty spots attracting many visitors including sandy beaches, coastal and valley fields abundant with natural wild life, where fish swim lazily in the clear water of the rivers. You can be sure of a warm friendly welcome from the locals.

In everlasting memory to the miners who lost their lives in Carmarthenshire.

Date	Mine	Lives Lost
10 May 1852	Gwendraeth	Twenty-six miners killed. Death came by water. The miners broke through into long abandoned workings, the youngest being ten years old.
7 March 1853	New Lodge	Collier Thomas William (12) killed by an explosion of firedamp in a tophole.
17 March 1860	Moreton	Collier James Morgan (35) killed by a fall of coal from the roof.
15 August 1860	Old Castle	Collier Walter Griffiths (39) killed by a fall of stone.
20 October 1860	Malton	Colliers Benjamin John (62), Thomas Griffiths (66) and Doorboy James Badham (11) killed by an explosion of gas caused by defective ventilation. The owner was summoned before the magistrates and fined.
31 October 1860	Dafen	Collier (15) killed when he fell into the shaft.
24 April 1868	Dyncwl	Labourer Jonah Williams (24) killed when he was struck by a windlass handle.
2 May 1868	California	Collier Elias Watkins (42) killed by a fall of roof.
27 May 1868	Glinea	Collier Evan Davies (42) killed by an explosion of gunpowder.
6 January 1872	California	Collier R. Roberts (23) killed by suffocation.
25 April 1872	Gors Coch	Haulier T. Samuel killed when he was crushed by drams.
8 July 1872	Old Castle	Trammer J. Thomas (23) killed by a fall of roof.
21 August 1872	Carway	Collier J. Jenkins (12) killed by a fall of coal.
17 December 1872	Bryn Gwyn	Collier D. Jones killed by a fall of stone.
9 January 1873	Woodbridge	Collier W. Davies (18) killed by an inundation of water from an old working.
7 April 1873	St George	Collier John Charles (43) killed by a truck on the surface.
6 May 1873	Clygwanen	Collier T. Jones killed by a fall of coal.
22 May 1873	Pencoed	Collier Wm Evans (26) killed by a fall of coal.
4 October 1873	Techon Fach	Collier John Jones killed by falling out of the cage whilst he was repairing the shaft.
9 October 1873	Pool	Labourer T. Rees (23) killed by falling on the surface whilst he was carrying timber to send down the pit.
14 February 1874	Capel Ifan	Collier J. Wilkins (22) killed by a fall of stone.
13 March 1874	Bryn Gwyn	Carter J. Samuel (15) killed by drams.
27 June 1874	Dynant	Sinker killed by falling from a bucket whilst ascending a sinking shaft
13 February 1875	Woodbridge	Ada Davies killed by getting entangled with the machinery which was improperly fenced.
8 May 1875	Kille	Collier Simon James (43) killed by a shot going off whilst engaged in unramming a hole which had missed fired. In doing so he was in violation of No.8 General rule.
15 July 1876	Cwm Capel	Collier D. Jones (25) killed by a fall of stone.
23 April 1884	Bon	Collier J. Emmanuel (48) killed by a fall of stone.
1 May 1884	Carway	Collier T. Williams (17) killed by falling under a dram.

Three
Pembrokeshire in the South Wales Coalfield

The Pembrokeshire Coalfield is assumed by various authorities to have been approximately seventy-six square miles. The fortunes of Pembrokeshire have always been linked with the sea; over the years many small ports and harbours have taken part in fishing and coastal trade and there were small ship building yards at places like Solva, Fishguard and Newport on the open coast and Dale, Angle, Burton and Cosheston on the more sheltered waters of Milford Haven. Wooden sloops and schooners were built just above high water mark by craftsmen using local timber and much local skill. Barges were also built for the limestone and coal trade. Sadly, by about 1870 most of the small shipyards had been forced to close.

The small Pembrokeshire Coalfield, part of the South Wales Coalfield, was made up of thin and disturbed seams of anthracite and, although the coal has been very difficult to work, there is a long history of mining going back to the fourteenth century. During the Industrial Revolution the coal was much in demand for its excellent heating properties. Many shallow bell pits, drifts, slants and shafts were opened up. Most of these were located in the Saundersfoot district and around the Daugleddau, but there were pits also near the shores of St Bride's Bay. Production reached its peak in 1885, with 100,000 tons of coal produced. Most of the coal was used locally, but there were exports from open loading beaches, small jetties and quays, and from Saundersfoot Harbour. But Pembrokeshire coal mining was always rather inefficient, and after a slow decline there were no pits left open after 1949.

The most important colliery district in the county was that around Kilgetty and Saundersfoot. Before 1900 there were several quite prosperous collieries at work, including the Bonville's Court Colliery, the Grove Colliery and Kilgetty Colliery (both at Stepaside), the Begelly Colliery and the Moreton Colliery. At first the coal was transported by horse and cart to the beaches at Wiseman's Bridge and Saundersfoot, where it was loaded onto sailing vessels at low tide. There were sometimes thirty vessels or more being loaded on Coppet Hall beach. Everything was changed after 1829 with the building of Saundersfoot Harbour and the various mineral railway lines. One line ran inland to Begelly and Thomas Chapel. The other branch ran northwards to the Stepaside area, where it served several collieries and the local iron industry. The line ran along the coast and needed several short tunnels, which are still in a good state of repair. Now coal could be exported easily and in the middle of the century the industry prospered. Several new collieries were opened up, and by 1864 over 30,000 tons of coal was being exported from Saundersfoot Harbour.

New industries, such as the firebrick works at Wiseman's Bridge, appeared on the scene. But as the best seams were worked out, the collieries closed one after another. By 1900 Bonville's Court was the only large colliery left open in the area.

Left: The remains of Grove Colliery, Stepaside, Pembrokeshire, in 1970. Grove Colliery was sunk in 1856 by Pembroke Coal & Iron Co. Grove Colliery was closed in 1903. *Right*: The remains of Kilgetty Colliery, Stepaside, in 1970. Kilgetty Colliery was sunk in the early nineteenth century to supply the Stepaside Iron Works. Kilgetty Colliery closed in 1939. In 1930 there were 130 abandoned mines listed in Pembrokeshire.

Testing for gas with a flame safety lamp in 1938. In the fourteenth century cats and dogs were lowered down the shallow bell pits and pit shafts by means of a rope for the detection of black damp (Carbon CO^2 and Nitrogen N^2), but mainly dogs were used. Animal cruelty was sometimes found, as some people just did not care for the animals' well being and safety. Domesticated rats and mice were also carried when canaries began to be used.

Bonville's Court Colliery, near Saundersfoot, in 1907. Bonville's Court Colliery was the largest colliery in Pembrokeshire and was opened in 1842 by Messrs Myers & Bonville's Court Colliery Co., who owned the concern since 1873. On 3 November 1911 an explosion occurred; fortunately there were no fatalities. Naked flame lights were in use. Bonville's Court Colliery was closed in 1930.

Hand boring machine in 1898. Approximate coal output produced from the Pembrokeshire Coalfield from 1875 to 1948: 1875, 80,000 tons; 1880, 82,000 tons; 1885, 100,000 tons; 1890, 89,000 tons; 1895, 85,000 tons; 1900, 50,000 tons; 1905, 50,000 tons; 1910, 45,000 tons; 1915, 55,000 tons; 1920, 55,000 tons; 1925, 50,000 tons; 1930, 30,000 tons; 1935, 50,000 tons; 1940, 30,000 tons; 1945-1948, 22,000 tons; 1948, 5,000 tons.

From the young miner to the gaffers, in 1894. By 1900 Bonville's Court was the only large colliery left open in the area. It survived until 1930, employing 300 men and producing over 30,000 tons of coal. Several other collieries opened during the twentieth century, at Reynalton, Broom, Kilgetty and Loveston. With the closing of the Broom and Kilgetty Collieries at the beginning of the Second World War the long industrial history of the Saundersfoot area came to an end.

Miners at work in 1928. Despite being among the most prolific in the country, the South Wales Collieries were to prove extremely difficult to mine. Work was always hard and fraught with danger and those who survived the inevitable explosions and roof falls became old before their time. Diseases like pneumoconiosis and silicosis almost invariably proved fatal and the eye disorder nystagmus, contracted from working at low light levels, could cause insanity if not treated.

Coppet Hall (Coalpit Hall) Tunnel, Saundersfoot in 2002. The three tunnels were completed around 1834 to carry coal from Stepaside Colliery, and mark the route of the old mineral railway line between Wiseman's Bridge and Saundersfoot. On the western side of the River Cleddau, the Hook district was much more successful and over the years many shallow workings, drifts and deep mines were opened. New Hook Colliery continued until 1948, sometimes producing more than 30,000 tons of coal a year.

Miners returning from work at Railway St (The Strand), Saundersfoot, on the Rosalind Express in 1901. Colliers working in small levels smoked cigarettes and pipes indicating that the place of work was thought to be 'more or less' gas free; many individual deaths and severe burnings did, however, occur in these small mines by small pockets of gas ignited by smoking and naked flame lights, which were used.

Coal loading bay at Saundersfoot Harbour in 1922. At first the coal was transported by horse and cart to the beaches at Saundersfoot, where it was loaded onto sailing vessels at low tide. Saundersfoot Harbour was built in 1829, and in 1864 over 30,000 tons of coal were being exported. The mines around the confluence of the western and eastern Cleddau Rivers also depended greatly upon the export of coal from small quays and jetties.

Early winding whim gin in the 1800s. A whim gin is a drum and rope used in early collieries for winding men, minerals and supplies etc. in the pit shaft. The axis of the drum was vertical and the drum was set at a height, which left room for the horse to walk a circular track around the drum, rotating it by means of a projecting beam to which the horse was attached. A horse pump based on the whim gin was also used underground for pumping mine water.

The remains of early bell pits at Templeton in the 1960s. A bell pit is an early coal pit, named because of its shape. At the bell pits near its crop in the vicinity of Templeton roof shales with fish-remains have been found, these are a characteristic feature of the spoil from the bell pits. The seam is said to vary from 2ft to 3ft of clean coal and to be very close to the surface. Bell pits were worked in Pembrokeshire from as early as the fourteenth century.

Thomas Chapel Wood Colliery in 1880. During the forty-six years before 1914, one miner was killed in South Wales on average every six hours. Causes were many and varied, but almost half the fatalities were down to roof falls. Explosions, although dramatic in the number of victims they claimed, accounted for fewer than seventeen per cent of the deaths. Sixty per cent of the victims were killed before reaching the age of thirty. Eighty per cent died before the age of forty. Thomas Chapel Wood Colliery was abandoned in 1938.

Left: Freystrop Colliery in 1906. The sinkers and materials were wound by means of a temporary flat winding rope. Freystrop Colliery was abandoned in 1912. One of the earliest areas, which shipped out coal in little sailing vessels, was the Landshipping area. Before 1844 there were five working collieries here, and several exporting quays. Landshipping village was a thriving mining community. The main quay was at Landshipping itself, and from here over 10,000 tons of coal and culm (small coal) were exported annually at the beginning of the nineteenth century. In 1844 water from the river broke into the Garden Pit Colliery workings, and more than forty men and boys were drowned. The mine was closed and the community never recovered. Many families emigrated and by 1867 all the other collieries in the area had closed. Further south, the collieries around Yerbeston, Cresselly and Jeffreston were also running down and after 1850 very little coal was exported from Cresswell Quay.

Below: West Park Pit, Hook Colliery, near the village of Hook, Llangwm. It was sunk in 1885 by Hook Colliery Co. to replace Green Pit, Old Aurora Pit, New Aurora Pit, Beam Pit and Commons Pit. In 1899 the colliery employed 103 miners. It closed for production in 1910 when New Hook, Margaret Pit was sunk. This winding engine had its rope drum mounted outside the engine house (centre of photograph).

The road dog naked flame candle light was the cause of many explosions in the pits in the nineteenth and early twentieth centuries, and some were made with a 'Dog Spike' (rail clamp). During the 1940s David Bennie & Sons supplied the mining industry with over 15,000,000 annually. The West Park Pit worked the very steeply inclined Rock and Timber Vein seams. West Park Pit, Hook Colliery was closed in 1910 by Watts, Watts & Co. Ltd. (*Wattstown, Rhondda Fach*).

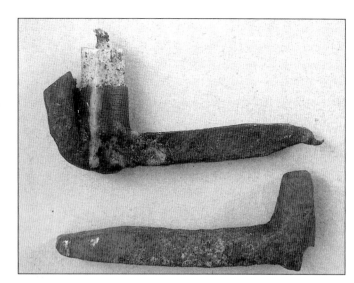

Old Hook Colliery, Llangwm, in 1912. Old Hook Colliery was opened in 1882 by Messrs Harcourt Powell. The seam worked was the Timber Vein. On 6 August 1902 at the Hook Colliery an explosion injured one man. Naked flame lights were in use. Old Hook Colliery was closed in 1910 by Watts, Watts & Co. Ltd.

Hook New Drift Mine, near Hook, Llangwm, in 1936. Sinking of an airshaft as part of the ventilation of a new drift mine at the Hook Colliery. The winding hand operated windlass is referred to as early as the eleventh century. The owners in 1936 were Watts, Watts & Co Ltd. In 1938 the colliery employed 113 miners. The mine was flooded by water from old working nearby which caused its closure. Hook New Drift Mine was closed in 1948 by the NCB.

Commencement of sinking the Margaret Pit, Hook Colliery, Llangwm, in 1907. It was known as New Hook Colliery. At this stage, men and materials were being lowered and raised in the shaft by the hand-operated winch. The St David's Peninsula is a place to visit; a land of pilgrimage, beauty and peace. Celtic, natural, spiritual: it's unique. A granite ledge of land jutting into the Atlantic, this western edge of Wales, is the patron saint's birthplace, his monastic settlement now a medieval cathedral at the core of Britain's smallest city. There's a heartbeat of the spirit here, tangible in monuments and moments. An important headland on sea routes linking ancient and Celtic cultures, many have passed this way. King Arthur landed on St David's shores. Black Bart, creator of the Jolly Roger, embarked on piracy from Solva harbour. Today seals, porpoises and dolphins inhabit the clear waters.

The Margaret Shaft, New Hook Colliery, in the process of sinking and reconstruction in 1907. New Hook Colliery was opened by New Hook Colliery Co. in 1910. In 1936 they sank the Hook New Drift Mine. New Hook Colliery was closed on Friday 23 April 1948 by the NCB. This is now a landscape with an elemental edge, all wrapped up in a clear, ocean-tinged light.

Men in a bowc during sinking of the Margaret Shaft in 1910. There were several small exporting quays in the area, for example at Little Milford, Hook Quay, Lower Hook Quay and Sprinkle Quay. After about 1850 cargo vessels became larger and more and more barges came into use for carrying the coal to Llangwm Pool and Lawrenny, where it was trans-shipped. But this was an expensive and time-consuming operation and after 1929 the Hook Colliery Railway exported most of the coal from Hook Colliery.

Women Navvies, Llangwm, in 1916. The women seen in the photograph are taking a break from their hard grafting day when they were building a donkey track to Little Milford. On St Bride's Bay, where there were sandy beaches close to the coal mining districts, it was difficult to build proper quays. Instead, the coal was exported in small vessels, which were beached on the sand at Nolton Haven and Little Haven. They were loaded amid hectic activity from horse-drawn carts at low tide and then floated off again on the next tide.

Trefrane Colliery, near Newgale. Trefrane Colliery, the most westerly colliery in the South Wales Coalfield. It was opened in the early nineteenth century by Walters & Canton. The Nolton and Newgale had six main collieries and many levels and slants; several levels can be seen in the coal seams in the cliffs between Newgale and Little Haven. Trefrane Colliery was closed in 1905.

Mighty industries come and go but Mother Nature ultimately prevails and we are left with the memories of human toil and the close knit communities which are its legacy.

In everlasting memory to the miners who lost their lives in Pembrokeshire.

Date	Mine	Lives Lost
14 February 1884	Landshipping	Thirty-seven miners killed. Those who died were: Thomas Gay, Benjamin Hart and his son, William Llewellyn and his son, Thomas Llewellyn, William Llewellyn, Benjamin Jones, Joseph Picton and his three sons, John Cole, Mr Hitchings, Mr Bedford, a boy named Thomas, a boy named Owens, two boys named Davies, two boys named John, two boys named Picton, a boy named Cole, a boy named Hughes, a boy named Hitchings, a boy named Llewellyn, a boy named Jones, a boy named Davies, a boy named Day, two boys named Butler, two boys named Cole, a boy named Jenkins and two orphans of the late Jane Wilkins.
14 November 1853	Little Milford	Collier's wife Mary Brock (52) killed by falling into a coal pit.
6 January 1860	Kilgetty	Trammer John Griffiths (13) killed by a fall of stone at the entrance to a stall.
17 March 1860	Moreton	Collier James Morgan (35) killed by a fall of coal from the roof.
3 April 1860	Hook	Trammer John Hier (15) killed when a bucket fell down the shaft on top of him.
5 September 1860	Freystrop	Collier Richard Allen killed by a fall of roof.
20 October 1860	Malton	Colliers Benjamin John (62), Thomas Griffiths (66) and doorboy James Badham (11) killed by an explosion of gas caused by defective ventilation. The owner was summoned before the magistrates and fined.
15 January 1875	Bonville's Court	Collier James Edwards (24) killed by a fall of roof.
2 October 1875	Hook	A pit overman was killed by falling down an air pit.
10 August 1885	Moreton	Collier John Wallace Thomas (20) killed by a fall of coal at the face owing to sufficient sprags not having been set. The roof also fell heavily and broke the timbering.
31 July 1886	Kilgetty	Furnace tender William Stephens (70) killed by being caught by the cage when entering after signalling it away. He was brought to the surface suspended by the neck between the shoe and guide rod. One leg was resting in the cage.
21 December 1891	Bonville's Court	Collier George Thomas (48) killed by a fall of roof 9ft x 3ft x 18in to nothing in thickness at the face of a heading 7ft 6in wide which was being driven through the seam of culm with a section of 4ft 6in. The roof which was shale was not double timbered up to the face. The Props were 3ft apart on each side of the roadway.
23 November 1892	Hook	Collier William Phillips (32) killed by a fall of roof at the face while he was finishing the lagging above a pair of timbers.
19 June 1906	Bonville's Court	Assistant trammer T.J. Hilling (14) killed.

Once again a sad reminder of the true price of coal. A sudden change; at God's command they fell; They had no chance to bid their friends farewell, Swift came the blast, without a warning given, And bid them haste to meet their God in Heaven.

South Wales Valleys

1   Thickly wooded, lush, green pastures,
Fishes filling pond and stream,
Birds of every size and colour,
Place of beauty, artist's dream,
Squirrels travelling down the Valley,
Leaping on from tree to tree,
Rivers flowing clear as crystal –
That's how our Valleys used to be.

2   Coal-mines sunk in every village,
People flocked from all parts then,
Row on row of hill-side houses,
Hooter's wail and black-faced men.
Strikes by miners followed later
When they struggled for their rights,
Rioting in Tonypandy,
Soldiers, marches, clashes, fights.

3   No work then, no money either,
Children standing each with bowl
By soup kitchens, thin and hungry,
Fathers couldn't claim the dole.
Jazz bands formed in every village,
Choirs' voices filled the air,
Chapels full, cymanfa ganu,
Neighbours all prepared to share.

4   'Bracchi' shops in every village,
Wafers, cornets, sweets galore,
Milk shakes, sundaes, cappuccinos
Served by Rosa or Aldó,
Fecci, Gazzi, Gambarini,
Coffee machines that hiss and steam,
A warm retreat for friends or lovers
To chat and laugh or weave a dream.

5   Tram cars, tin baths, silicosis,
Trips to Barry and Porthcawl,
Greyhounds, pigeon-cotes, allotments,
Dramas in the Workmen's Hall,
Brass bands, lodgers, penny readings,
Boxing, rugby, fish and chips,
Dai-capped, mufflered, blue-scarred miners,
Pit explosions, slag-heaped tips.

6   King Coal's sovereignty has ended,
Only Tower still remains,
And the links have now been broken
In the Valleys' pit – life chains.
But the courage of coalminers,
Their brave spirit, toil and sweat,
Their proud struggles, pain and hardships
And sufferings we must not forget.

Hawys Glyn James